Post-Harvest Insect Disease Management

Post-Harvest Insect Disease Management

Ruchika Sekh
Nikhil Bano
Bhavesh Arya

2025
Agri Horti Press
New Delhi-110 084

ISBN: 978-93-9296-926-3 (Hardbound)

Published By

Agri Horti Press

New Delhi

E-mail: agrihortipress@gmail.com

Preface

"Post-Harvest Insect Disease Management" serves as an indispensable guide to addressing one of the most critical challenges in agricultural production and food security—protecting harvested crops from insect pests and diseases. This book offers a comprehensive overview of strategies, techniques, and technologies for managing post-harvest losses due to insect infestations and diseases, providing valuable insights for researchers, practitioners, policymakers, and stakeholders involved in the agricultural supply chain.

Post-harvest losses caused by insect pests and diseases pose a significant threat to global food security, jeopardizing the livelihoods of millions of farmers and contributing to economic losses and food waste. Effective management of post-harvest pests and diseases is essential for ensuring food quality and safety, reducing losses, and enhancing the efficiency and sustainability of the food supply chain.

The primary objective of this book is to provide readers with a comprehensive understanding of post-harvest insect disease management, encompassing key topics such as insect biology and behavior, disease dynamics, monitoring and detection techniques, integrated pest management (IPM) strategies, and innovative technologies for pest control and preservation.

Contributions from leading experts in the field cover a wide range of crops and commodities, including grains, fruits, vegetables, nuts, and stored products, as well as various storage and processing facilities. Special emphasis is placed on the principles of IPM, which advocate for the use of multiple strategies, including cultural, biological, and chemical control methods, to minimize the use of pesticides and mitigate the risk of resistance development.

As we navigate the complexities of post-harvest insect disease management, it is crucial to adopt a holistic and multidisciplinary approach that considers ecological, economic, and social factors. By integrating scientific knowledge, technological innovations, and stakeholder engagement into pest management strategies, we can develop sustainable solutions that protect food quality and safety while safeguarding environmental and human health.

We extend our sincere gratitude to the contributors whose expertise and dedication have enriched this book. It is our hope that "Post-Harvest Insect Disease Management" will serve as a valuable resource and practical guide for researchers, practitioners, and policymakers striving to mitigate post-harvest losses and ensure a more resilient and sustainable food supply chain for future generations.

Ruchika Sekh

Nikhil Bano

Bhavesh Arya

Contents

1

POSTHARVEST HANDLING

Vegetables and fruit are living organisms that continue to change after harvest. While some of these changes are desirable, most are not, and growers must be aware of effective ways to minimize undesirable changes, increase shelf life, minimize food safety risks, and decrease postharvest losses. For most vegetables, maintaining cool temperatures and high humidity are the most effective means of preserving quality.

Once picked, vegetables will respire, meaning they use their stored sugars to produce carbon dioxide and heat. The more rapid the rate of respiration, the faster a vegetable will use up the stored food supply; the greater the heat produced, the shorter the postharvest life of a given commodity.

Vegetables also give off ethylene, a ripening hormone which promotes senescence. Detrimental effects of senescence include loss of green color; abscission of leaves or flowers; toughening of asparagus spears; russet spotting in lettuce; sprouting of potatoes; bitterness in carrots; and general weakening of the vegetable, which greatly reduces the natural resistance to decay organisms. The effect of ethylene is influenced by the amount present, the length of time the vegetable is exposed, and the temperature. Exposure to a specific concentration of ethylene for a given time will have much less influence at 32°F than at 85°F. The sensitivity of many vegetables to ethylene increases with maturity or age.

Transpiration, the loss of moisture from living produce, is one of the primary determinants of postharvest life and quality. The rate of moisture loss depends on both the commodity and the environment and is influenced by many physical and morphological factors. These factors include storage environment, surface to mass ratio (e.g., leaf lettuce has much more surface area per weight than winter squash and is more subject to weight loss), and injury. High humidity also helps to limit moisture losses. See specific crop chapters for best storage temperature and humidity recommendations.

HARVEST CONSIDERATIONS

Harvesting tools, equipment, and containers, must be cleaned and sanitized, when possible, before harvest begins and anytime they become dirty. Cleaning tools, detergents, and sanitizers must be provided so sanitation practices can be completed. Workers and visitors who contact vegetables and/or food contact surfaces, also must have clean hands. Well-stocked and clean toilet and handwashing facilities must be provided to all employees and any visitors. Employees must be trained on how to properly wash their hands as well as when handwashing is critical, such as after using the toilet, after eating, and anytime they may be contaminated due to contact with animal or other sources of illness causing organisms.

Clean Surfaces and Containers

Dirty surfaces can also transmit decay and illness causing organisms, Ensuring harvest and postharvest tools and containers are clean and sanitized prior to use will reduce both postharvest losses and food safety risks. If new containers are used, ensure they are stored in clean areas prior to use to prevent contamination. The concern about clean surfaces extends to the hands of those involved in post-harvest handling such as those who cull and pack produce. Proper hand hygiene including handwashing will reduce food safety risks that could be introduced during postharvest handling.

Washing and Chlorination

Decay is usually the most obvious postharvest problem but food safety risks should also be a consideration. Many decay and illness-causing organisms (bacteria and fungi) cannot invade sound, undamaged tissue, but as the tissue becomes older, it becomes weaker and more subject to invasion. To control postharvest losses and reduce food safety risks, it is recommended that produce be washed in chlorinated water before storage or shipping . The wash temperature should be about 10°F warmer than the produce temperature to ensure that decay organisms are not sucked into the tissue. For a 25 parts per million (ppm) solution, use one tablespoon household bleach (5.25% sodium hypochlorite) per eight gallons of water or one pint bleach per 264 gallons. Since chlorine is most effective at a slightly acidic pH, it is important that wash water is buffered to adjust the pH to between 6 and 7.

Chlorine in the wash water is often inactivated when the wash water becomes dirty. Use filtering devices to remove soil and organic material, and check the chlorine concentration often. Produce should be subjected to the chlorinated wash from one to ten minutes. After it is removed, allow it to drain for several minutes before packing. There are other chemicals beside chlorine that can be used such as those that are peroxide or peroxyacetic acid based. Whatever is used, it should be labeled for use on vegetables.

Amount of sodium hypochlorite to add to wash water for 50-150 PPM dilution.

TARGET PPM	ML/L	TSP/ 5 GAL	CUP/ 50 GAL
Sodium Hypochlorite, 5.25%			
50	1.0	3 2/3	3/4
75	1.4	5 1/2	1
100	1.9	7 1/4	1 1/2
125	2.4	9	2
150	2.9	11	2 1/4
Sodium Hypochlorite, 12.75%			
50	0.4	1 1/2	1/3
75	0.6	2 1/4	1/2
100	0.8	3	3/5
125	1.0	3 3/4	4/5
150	1.2	4 1/2	1

Cooling

Cooling is important for most commodities, and a quick cool down or precooling is desirable for highly perishable products that are harvested or packed in warm weather. In general, a vegetable that has a storage life of 30 days at 32°F will have a storage life of about one to three days at 90°F or three to six days at 70°F. Precooling methods are usually dictated by the product and the available facilities or resources.

Chilling Injury

Chilling injury is caused by low, nonfreezing temperatures and generally affects vegetables with a higher recommended storage temperature. See individual crop chapters. Symptoms of this disorder include decay; failure to ripen properly or uniformly; pitting of the surface; discoloration (russetting on the surface or darkening of the flesh); and watery consistency. Not all symptoms are obvious, and several vegetables have specific symptoms that are not described. The injury incurred is dependent on time and temperature. The lower the temperature below the recommendation, the quicker and more severe the injury. Chilling injury is particularly troublesome because symptoms do not appear until after the injury has occurred. Produce that looks healthy in the cooler may develop symptoms during transit or marketing that were caused by chilling before it left the farm.

POST-HARVEST INSECT PEST MANAGEMENT IN SMALL FRUIT CROPS

Grape

In general, it is advisable to maintain healthy foliage as long as possible into the fall. Here are some insect problems that can be addressed to help achieve this goal. Grape mealybug and some related mealybug species can be found feeding on grapevines during much of the year. The phloem-feeding insects are covered with white, waxy material that often is produced into finger-like projections along the sides of the body, and into long white filaments from the hind end (all of these disappear if an impressive looking mealybug is dropped into a vial of alcohol!). Grape mealybug is a pest that can be addressed either early in the year (the recommended timing in our southeastern regional recommendations, https://smallfruits.org/ipm-production-guides/), or in the fall. Traditionally, mealybugs should be controlled if there were high populations in the fall – honeydew and sooty mold can cover foliage and clusters, and high populations can cause clusters to drop. The situation has been made worse by the advent of grapevine leafroll virus which is transmitted by mealybugs. Mealybugs can be targets of natural enemies, but broad-spectrum sprays can induce outbreaks of these insects. Venom, Assail, Actara (thiamethoxam), Admire Pro (imidacloprid), Belay (clothianidin), Applaud (buprofezin) and Movento (spirotetramat) are good candidates for chemical control of mealybugs. All but the last two are neonicotinoids.

Fig. 1. 1 A young grape mealybug on a Virginia grapevine.

The question occasionally arises about the utility of oil sprays to grapevines in the dormant period – these have long been used in orchards. A usual target for oil sprays is European red mite, which overwinters as eggs on the bark of fruit trees and grapevines. On the relatively smooth bark of apple and peach branches, a coating of oil acts to suffocate the eggs (especially if applied close to hatch, when oxygen needs are greatest). However, the shaggy bark of grapevine trunks makes it difficult to achieve adequate coverage of the eggs with an oil spray. Another insect that may be targeted by oil sprays is grapevine scale. Vines may develop populations of scales especially if broad spectrum sprays have suppressed populations of parasitoids. On tree fruits, aphids are also the targets of oil sprays since they overwinter as eggs on the bark. But the main aphid on grape uses black haw as its winter host, so aphid management in vineyards does not benefit from oil sprays.

Fig. 1.2. Grapevine scales on a Virginia grapevine. The large individuals are the scale coverings of mature females, and the small, white individuals are settled crawlers.

General examination of vines. Vines may be examined in the post-harvest period for symptoms of Pierce's disease, including typical leaf discoloration and the presence of matchstick petioles. Unusual pest situations may also be revealed, such as lecanium scales. European fruit lecanium and terrapin scape could occur on grapevines but this is not common.

Caneberry

Rednecked cane borer is a member of the flatheaded borer family Buprestidae. The elytra are black or dark grey, and with a distinctive copper-colored pronotum. The adult beetle is about a quarter inch long. Adults are around for a good part of the season, from May to August. Larvae make spiral tunnels under the bark, reducing growth of the plant. When sprays are made for this insect, the target is ovipositing females. During winter pruning, assess the presence of galls. If more than 10% of canes have galls, a chemical control approach is needed in the coming season. But pruning during the dormant period can help reduce populations. Removal via pruning is most effective if wild brambles growing nearby are also removed. Removal of galled canes can be helpful at other times of the season as well.

Fig. 1.3 Rednecked cane borer adult and galls. Susan Ellis, James Solomon, Bugwood.org.

Raspberry crown borer is a clearwing moth (Sesiidae). Moths in this family are often wasp mimics, and this one is a good mimic of a yellowjacket. Crown borers invade the crown of the plant where they can cause girdling. After the eggs hatch, the larvae spend the winter in hibernacula beneath the soil surface. During the first summer, larvae feed within the crown; in the second summer, they tunnel upward into canes. They may be controlled by a soil drench in the fall, while they are preparing for overwintering, or in April, before they tunnel into the plant. Brigade (bifenthrin) or Altacor (chlorantraniliprole) may be used, applied in a minimum of 50 gallons of water per acre. In addition, infested plant may be rogued out.

Fig. 1.4 Raspberry crown borer and larvae in its gallery. Univ. Georgia, Bugwood.org.

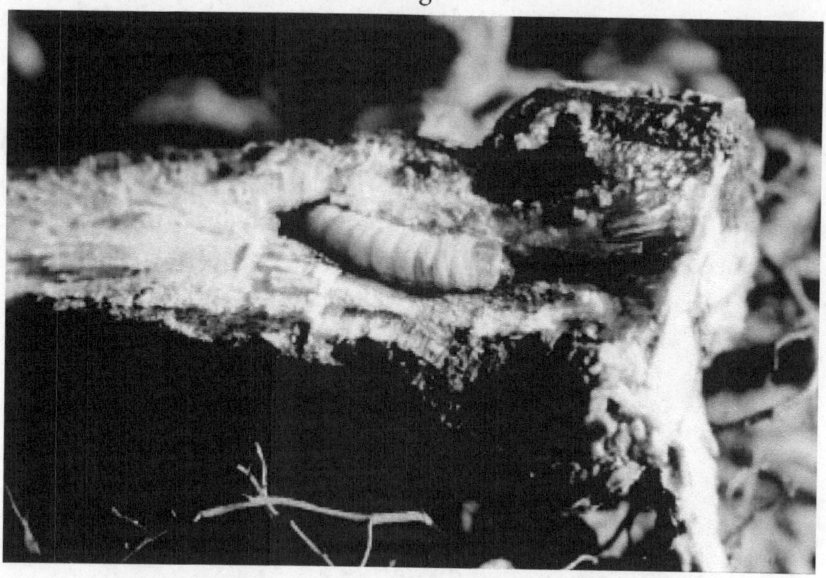

Fig. 1.4.b Raspberry crown borer and larvae in its gallery. Univ. Georgia, Bugwood.org.

Rose scale may occasionally develop on caneberries. Rose scale overwinters as eggs beneath the maternal scale covering. There are 2-3 generations in the north, but more in the south. This scale can be controlled during the dormant or delayed dormant period using Admire Pro (imidacloprid), Brigade (bifenthrin) or a 2% solution of Tri-Tek oil. Sprays should be made when temperatures are above 50 degrees F.

Fig. 1.5. Rose scale infestation in a caneberry planting in Blackstone Virginia.

If high populations of aphids, mites or leafhoppers are present after harvest, they can be controlled as during the pre-harvest period.

Strawberry

Since strawberries are harvested early, for perennial or matted-row crops much of the remaining season is a post-harvest season. Foliar feeding pests can be controlled as normal. A couple of pests bear specific mention.

Strawberry root weevil mainly occurs in beds that are in place for several years, but not in beds that have a shorter lifespan. The adults (all females) are flightless, and since they have to walk into the planting, and only have a single generation, populations need some time to build. If root weevils need to be controlled, Brigade (bifenthrin) or malathion may be applied when leaf-feeding by adults appears.

Fig. 1.6. Strawberry root weevil adult. Whitney Cranshaw, Bugwood.org

White grubs may be a problem in strawberries, and Admire Pro can be applied at renovation, when it is incorporated into the soil.

Blueberry

Blueberry bud mite is a tiny mite in the rust or blister mite family (Eriophyidae). Mites in this family tend to be highly host specific, and blueberry is the only host. The mite invades flower buds in late summer and fall, where they feed during the winter. Attacked buds fail to expand and bloom, or give rise to malformed, roughened fruit. The biology of this pest is not well known. Treatments may be considered in blocks of susceptible varieties with a history of infestation. High volume and high-pressure applications of horticultural oil may be made. Summer topping and hedging right after harvest helps remove older, infested tissue and can help in blueberry bud mite management.

Harvest Stored Product Insects and their Management`

1. Introduction

Advancement of technologies in agriculture has led to increased food production every year. In many countries, major portion of the food grains produced, were storing for contingency and regular supply.These stored grains were infested by insects directly or indirectly and causing severe damages, apart from other storage losses. Most of the storage part is mainly concentrating on grain storage either in domestic or commercial scale. Several structures were used for grains storage ranging from a small metal bin to tall grain elevators. Such stored commodities for long time were prone to contamination and damage by biotic and abiotic factors. Among the biotic agents, insects, mites, rodents, birds and microorganisms cause immense loss in storage. Major damage was mainly caused by insects, which account for an average of 10-20 per cent of storage losses (Phillips and Throne 2010). In general, stored products of agricultural and animal origin are attacked by more than 600 species of coleopterans, 70 species of lepidopterans and about 355 species of mites. They are causing both quantitative and qualitative losses (Rajendran and Sriranjini 2008).

The insects damaging the stored grains, many times got access from the field and established at the storage site due to microclimate and retained during the course of processing and storage (Hagstrum and Phillips 2017). Reports have shown that some insect pests initiate damage during ripening stage of crops and continue to storage. However, major sources of infestations are old bags, storage structure, old containers and cross over infestation (Perez-Mendoza et al. 2004), harvesters and other machineries (Sinclair andWhite 1980).The initial infestation can be minimized during the post-harvest handling to the storage structures by proper harvesting and drying of grains. However, between the lots the movement of stored product insects are facilitated by grains supply from one region to another either through the commodity. Sometimes they also spread by active flight of insect pests as many of them are strong fliers (Mahroof et al. 2010; Ridley et al. 2011a, b). Nearly 100 species of insect pests of stored products lead to economic losses. It is estimated that in India, crop loss due to pests ranges from 10-30 per cent a year, out of which 26 % is due to the insect pests.

Storing of food grains throughout the World is an age-old practice. The period of storage from harvesting to the consumption of grains is either seasonal for seed purpose or over the years for buffer stock. During storage, the grains undergo several changes due to physico-chemical and biological grain metabolism which led to their deterioration. The physico-chemical changes may occur due to abiotic factors like temperature, carbon dioxide, oxygen and moisture level, while biotic factors included fungi, bacteria, insects, mites, rodents and birds.

Stored-grain insect-pests may cause as much damage during storage as crop insect-pests cause during the growing season. Insects can damage the grains by direct feeding or indirectly by deterioration and contamination of grain by their faecal matter, exuviae or secondary infestation. Direct feeding damage results in reductions in grain weight, nutritional value, germination and market value. Whereas, deterioration and contamination from the presence of insects results in downgrading of grain and market value due to insect parts, odours, molds and heat damage.

The quantity so wasted is enough to feed at least 50 million people. This loss is not merely in terms of quantity but also in terms of quality of food grains. Many insect species are associated with stored grains and their products, but only about 50 are severe, either occasionally or frequently.The relative abundance of particular species may vary from country to country. Majority of them are distributed throughout the world and were reported to cause similar type of damage everywhere, although damage severity varies. In many cases, complexes of the insect species are present in the commodity lot as their feeding niche is different and can sustain simultaneously. The particular insect present also depend on the type of the stored grain and region of warehouse.

CLASSIFICATION

The commonest classification of the storage insects is based on their feeding habits as 'primary pests' and 'secondary pests'. Primary feeders are those, which are able to damage the whole, undamaged grains. Their lifecycle involved with the whole grain as their young ones bore and feed inside the kernel part of the grain which sometimes creates 'hidden' infestation. These can cause very severe damage to the lot and if unnoticed till their population establishment, they are hard to manage. Regular monitoring is the essential step for preventing their damage. Secondary feeders are generally called as 'bran bugs' because they establish on the grains, which were already damaged either by the primary pests or other miscellaneous damages. They generally survive on broken kernels, debris or higher moisture weed seeds. The immature stages of these insects were found external to the grains and generally cannot initiate the infestation of the whole grains. But once after establishment these are generally contributing to spoilage, but not as severe as primary insects do. Their damage can be easily identified as their life stages were mobile in the commodity area. Some of the external pests are also mold/ fungal feeders. They contaminate the grains through their presence and metabolic wastes, which generates moisture (excretion and condensed heat) for mold development (Magan et al. 2003). Thus, the mold development indicates the final condition of grain spoilage.

Grain weevils

a. Rice Weevil: Sitophilus oryzae (L.) (Coleoptera: Curculionidae)

Rice weevils are prolific breeders and can build up huge populations in stored grains to the point where the grain has little value as a food product. During heavy infestations, heat and moisture are produced, leading to colonization by molds and mites. Attacks from Sitophillus oryzae can start in the field, when the moisture content is about 20%.

Egg

- Translucent white, plugs the egg hole with gelatinous secretion, laid singly on grains.
- Female weevils will deposit 300- 575 eggs in grains over a 6-8 month life span.
- Eggs hatch in about three days.

Adult

- These small
- beetles are reddish brown, and about 2.5 mm long and have 4 yellowish patches.
- They have distin- ctively long
- snouts and numerous round pits on the thorax.

- These weevils are able to fly.

Pupae

- The pupa is naked and the pupal stage lasts an average of 6 days.
- Pupates inside the grain.
- Larvae
- The immature stage is approxim- ately 2-3 mm long, white to yellowish in colour.
- It has a grub like in appearance and has no visible legs.

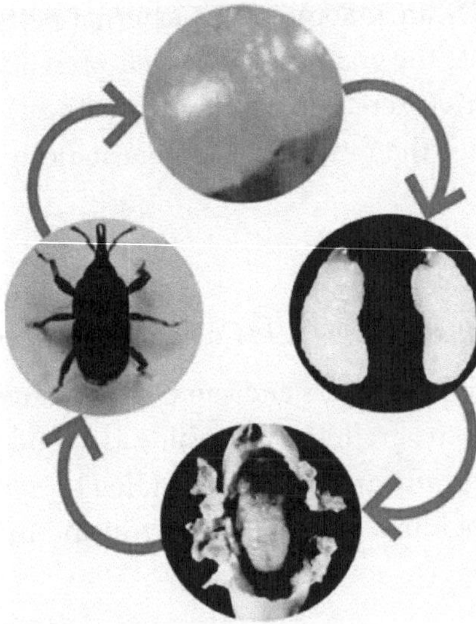

Fig. 1.7 Rice Weevil: Sitophilus oryzae (L.) (Coleoptera: Curculionidae)

- Distribution and status: World-wide and is found practically throughout India. It is the most destructive pest of stored grain. The rice weevil may be found in the paddy fields as well.
- Host range: Wheat, oats, rye, barley, rice, and maize.
- Damaging stage: Both the adult and larva (grub) Symptoms of damage:

The adults attack whole grains. Both adults and grubs feed inside grains, leaving large cavities and emergence holes.The adults also attack damaged grains.

Detection:

The eggs, larvae and pupae are not normally seen because they develop inside intact grains. The larvae chew large, irregular holes in the germ and endosperm of the kernel. Adult emergence holes (about 1.5 mm diameter) with irregular edges are apparent some weeks after the initial attack. Two rice weevils may develop at the same time on two sides of a single kernel. Adults can be found wandering over the surface of grain. In a heavy infestation, the only part of a grain that remains is the shell of the kernel perforated by adult feeding and emergence holes.MaleS. Orzyaeproduce an aggregation pheromone ((4S, 5R)-5-Hydroxy-4-methylheptan-3-one) to which males and females are drawn.

b. Granary weevil: Sitophilus granarius (L.)(Coleoptera : Curculionidae)

Granary weevil is a small, moderately polished.The insect resembling the rice weevil, and is commonly confused. It prefers a temperate climate. Both the adults and larvae feed voraciously on a great variety of grains. Female make a scoop in grain and then deposits an egg and covers it with a gelatinous fluid. In warm weather, the granary weevil develops from the egg to the adult stagein about 4 weeks. Cold weather greatly prolongs the developmental period.The life cycle is similar to rice weevil.

c. Maize weevil: Sitophilus zeamais (Motschulsky) (Coleoptera : Curculionidae)

It distributed in tropical and temperate areas, warm humid areas where corn is grown but can be found in colder climates like Canada also.The weevil having a major host range of maize, rice, sorghum, wheat, casssava and yam, and also dried stored products, including pasta. Both the grubs and adults are damaging stage. The life cycle is similar to rice weevil.

Grain borers

a. Lesser grain borer:
Rhyzopertha dominica (Fabricius)(Coleoptera :Bostrichidae)

The lesser grain borers a commonest grain boring beetle in the country. It is small and was also commonly known as wheat borer.The insect can be easily distinguished by its reflexed head, which is a typical character of its family. This beetle is tropically originated and most commonly encountered in India in cereal grains stored under temperature range of 20-30 °C, can able to survive 47°C.

Egg

- The female lays her eggs one at a time or in batches of up to 30 and can lay up to 500 eggs over 3 or more months.

Adult

- The adult beetle is dark reddish- brown with a cylindrical body about 3 mm long.

- The adult may live up to 240 days and is a strong flier.

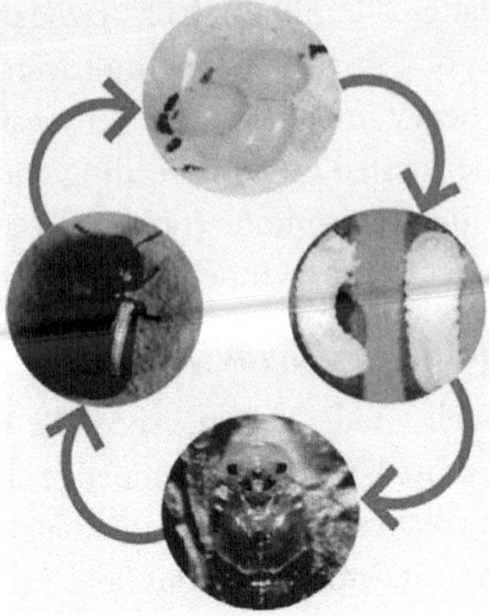

Fig. 1.8 Lesser grain borer, Rhyzopertha dominica (Fabricius)(Coleoptera : Bostrichidae)

Pupae

- Grub enters the grain after 3rd instar for pupation.

Larvae

- The larva is white, and as it matures, it becomes C- shaped and immobile.
- White, apodous with brown head, free living upto 3rd instars.
- Distribution and status: India, Algeria, Greece, United States, New South Wales (Australia), Japan, China.
- Host range: Paddy, rice, wheat, maize.
- Damaging stage: Both adults and larvae. Symptom of damage

Both adults and grubs bore the grains and feed voraciously. Severe infestation leads to only leave with frays. Insects are not swift movers and can easily hide in the cracks and underneath damage flour.

Detection: The insects can be either detected based on the random sampling and record for the holes in the grains or by adults moving on grain. Grubs and adults completely feed on internal material and leave the husks and flour with sweet smell/fungus smell at severity. The insect outside the grain can be detected by sieving which discharges the adults to fall down from the grain.

Morphologically adults are dark reddish brown in colour. Female insects lay eggs in single or in groups. Grubs are whitish and are funneling inside the grains.

b. Pulse beetle: Callosobruchus chinensis (L.) (Coleoptera:Bruchidae)

Callosobruchus spp. is important primary pests of pulses. Infestation may start in the pods before harvest and carry over into storage where substantial losses may occur. Levels of infestation may be high.The insects are able to cause damage in almost all types of stored pulses.Theyprefer whole grains for damage initiation/egg laying.

- Distribution: Cosmopolitan.
- Host range: All whole pulses, beans and grams.

Egg

- Egg period is 6 -16 days
- Fresh eggs are translucent, orange cream in colour, changing to greyish white with age.

Adult

- The adult beetle measuring 3-4 mm in length
- The average life- span of an adult is 5-20 days.

Pupae

- The pupal stage lasts 4-28 days.

Larvae

- Larva is whitish with a light-brown head.
- The mature larva is 6-7 mm long.
- Larval period 10 - 38 days.

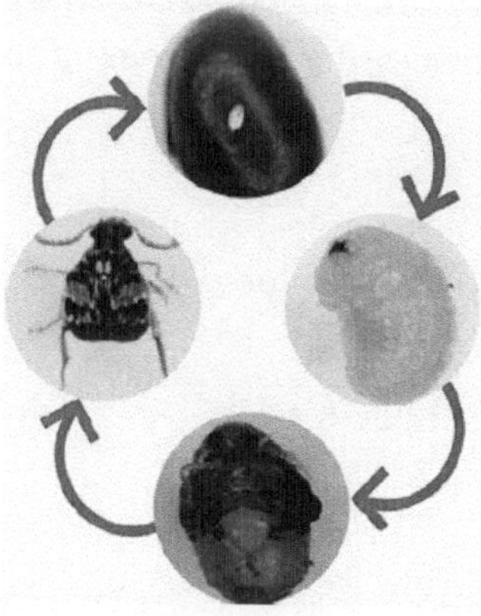

Fig.1.9 Pulse beetle, Callosobruchus chinensis (L.) (Coleoptera:Bruchidae)

- Damaging stage: Both adults and larvae. Symptom of damage:

Female insects lay eggs on the whole grains and this is the initial symptom or early detection method of insect. The grub after hatching enters into the seed and devours completely the internal material and leave the empty outer coat.

Detection: Callosobruchus spp. Lays egg on the grain surface, this presence of eggs is the first lineof detection. After feeding of the grain the exit holes caused by the adults was one more detection symptom. But by the time exit holes are visible, the insect completely infested the grain and of no use.

Pulse beetles are generally got the name because they mainly feed on all pulses. The insects are dynamic and fly quickly at exposure to light. Although seven species of Callosobruchus are common in india, C. chinensis, C. maculates, C. analis are severe.They are generally differentiated based on the colouring pattern of body. Adults are short lived and their main purpose is to find mate and lay eggs for the continuation of generation.

5. Grain and flour beetles

(a) Saw toothed grain beetle: Oryzaephilus surinamensis, (Coleoptera: Bruchidae: Silvanidae)

Saw toothed grain beetle is one of the best known cosmopolitan species. It is a slender, flat brown beetle. The thorax part has six saw tooth like projections, hence the name given. These insects are recently placed in silvanidae, earlier are under cucujidae.These insects are typically secondary pests mainly feed on damaged grains primarily damaged by other insects or mechanical damage. Larva can attack the germ in whole cereal grains. They reduce the nutritional content of the grains and reduce germination. In addition, damage grains produce off odour upon severe infestation.Though the insects are winged but rarely fly. They always wandering in search of food and rest in crevices, ducts and roofing spaces, from they are difficult to eradicate.

Egg

- Eggs are white in colour, laid loosely in cracks of storage receptacles and godowns.
- Hatching Period is 3-17 days.

Adult

- Adult insects are narrow, flattened having six teeth like serrations on each side.

- Female can lay upto 300 eggs.

Larvae

- Larvae are slender with two darker patches on each segment.
- Larval period is 14- 20 days.

Pupae

- Full grown larvae form the cocoon like covering with sticky secretion and pupate inside.
- Pupal period is 7-21 days.

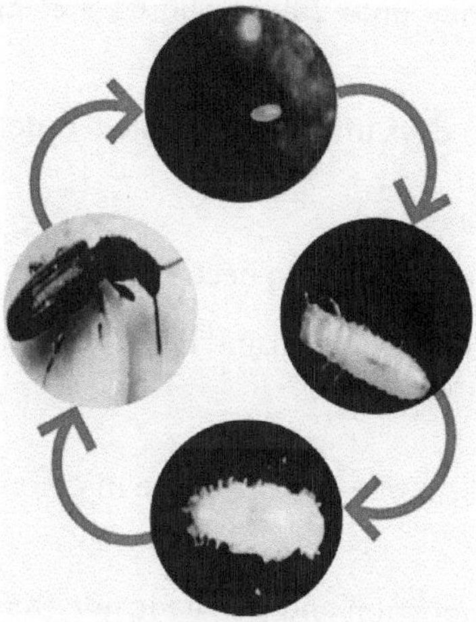

Fig. 1.10 Saw toothed grain beetle, Oryzaephilus surinamensis (Bruchidae:Silvanidae)

- Distribution and status: Cosmopolitan and common in worldwide pest of grain and grain products as well as chocolate, drugs and tobacco.
- Host range: Feed on food of vegetable origin, grains and grain products
- Damaging stage: Both adults and grubs Detection:

Adults can be seen upon careful observations, if the number is more. However, insects are able to climb on glass surface also; they can be seen on container walls. Using pitfall traps for detection of these insects is unsuitable idea. Since these are secondary pests, generally they found in the damaged grains.

(b) Rusty grain beetle: Cryptolestes ferrugineus (Stephens) (Coleoptera : Laemophloeidae)

These insects are reddish brown in colour, most common instored cereal especially wheat. Adults and larvae feed mostly on the germ portion of grains.

Severe infestation causes increase in grain temperature and contaminate the grains with faeces and also spread the fungal spores.Though the adults are winged but rarely fly. Adult insects walk with a characteristic swaying movement.

- Distribution and status: Worldwide distribution and is resistant to cold than other flat grain beetles.
- Host range: cereal grains, oilseed cakes, tobacco and dried store products
- Damaging stage: Both adults and grubs

Egg

- Each female can lay upto 200 to 500 eggs & Eggs are laid loosely on or among the grains.
- Hatching in 3 to 5 days under favourable temperature (30°C).

Adult

- Adult is a shiny reddish brown beetle about 2 mm long.
- It moves rapidly in warm grain and lies within the temperature above 25°C.

Larvae

- The larvae are worm like and are white in colour with two project- ions at tail end.
- Feeding on germ portion and pupate inside the grain.

Pupae

- Pupation is inside the grains after feeding the germ portion.
- The total lifecycle ranges from 35 to 150 days based on the climatic condition.

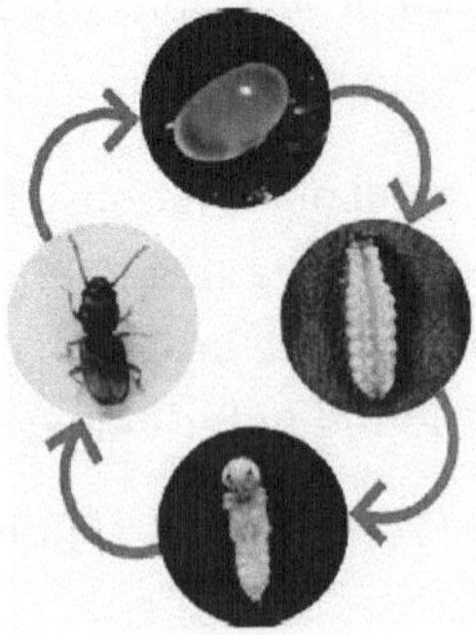

Fig. 1.11 Rusty grain beetle: Cryptolestes ferrugineus (Stephens) (Coleoptera: Laemophloeidae)

Detection:

These are several closely related Cryptolestes spp. with similar appearance and habits. Larvae and adults characteristically feeding externally on damaged grains.

(c) Rust-red flour beetle: Tribolium castaneum (Herbst) (Coleoptera :Tenebrionidae)

Red rust flour beetle and confused flour beetle are two almost similar insects generally confusing between. These beetles are generally encountered with stored food products irrespective of whether primary and processed ones. These are secondary invaders which are not able to cause primary damage and associated with the primary feeders. They multiply rapidly when the food source is rich with protein material. They can easily establish in the households, granaries, shipment mills, warehouses etc.

Egg

- White, translucent, sticky, slender and cylindrical.

Adult

- Oblong,flat, brown in colour.

Pupae

- The pupa is yellowish and hairy.
- The pupal stage lasts 5-9 days.

Larvae

- The young larva is yellowish white and measures 1 mm in length.

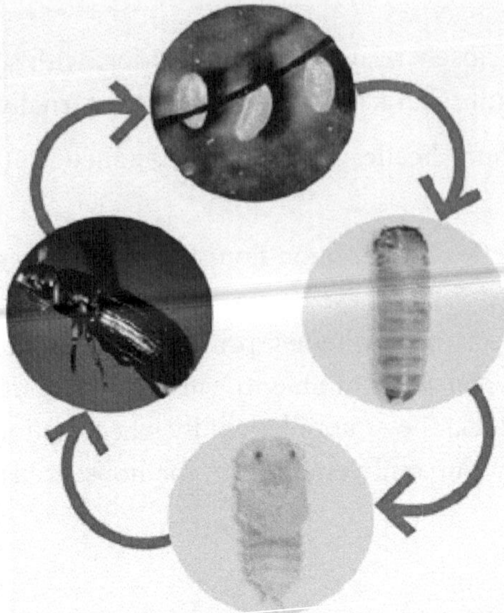

Fig. 1.12 Rust-red flour beetle:Tribolium castaneum (Herbst) (Coleoptera : Tenebrionidae)

- Distribution and status: Cosmopolitan, but more common in warmer regions. This pest occurs in temperate areas, where it survives winters in protected places, especially with central heating.
- Host range:Wheat-flour, dry fruits, pulses and prepared cereal foods, such as cornflakes.
- Damaging stage: Both adults and larvae. Symptom of damage:

Adults and larvae feed mainly on the germ of the cereal. Detection:

At lower densities, these insects are difficult to detect. However, in flour tunnels can be visible because of the adults and grubs movement. The flour kept for 15-20 days undisturbed generally infested by these beetles and bad odour is a symbol of severe damage. The grain/flour quality completely deteriorate, if the measures not taken up immediately. The adult are fast movers and good fliers and are easily visible, is the main sight of detection.

Adults are reddish brown i.e. rusty red and the body is flattened. During unavailability of food these insects overwintersin cracks and crevices of the warehouses/walls. The grubs and pupae are pale in colour.These insects can be easily separate from flour by sieving.They are the main host for many laboratory studies. In India, this insect is reported with phosphine resistance upto 35 folds.

(d) Flat grain beetle: Cryptolestes pusillus (Schönherr)(Coleoptera :Laemophloeidae)

Flat grain beetle belongs to Cryptolestes genus are minute pests of stored grains. Cryptolestes pusillus and C. minutes are the two important species commonly found. Recently the group is included and pronounced as Laemophloeus belonging to Laemophlocidae family. These are smallest beetles, both adults and larvae attack almost all types of grains and grain products including dried fruits, groundnut seeds and oilseed cakes. Rapid reproductions leads to increase in number and are generally found in association with many other stored insects like rice weevil, flour beetles etc.

Egg

- Female lays eggs in seed/grain cracks or even drop them loosely upon the farinaceous material.
- Eggs are white in colour and hatch within 5-6 days.

Adult

- Adults are reddish brown beetles smallest of all stored grain insects, approximately

1.5 to 2.0 mm.

- The total life cycle is 40-46 days.

Larvae

- Larvae are gene- rally active in feeding and after hatching, they search the damaged grains. The larvae also feed on dead insects.
- Larvae are cigar like yellowish white having the larval period of 21 days.

Pupae

- Fully grown larvae form like gelatinous cocoons adhering to the food materials, pupation is inside these cocoons.

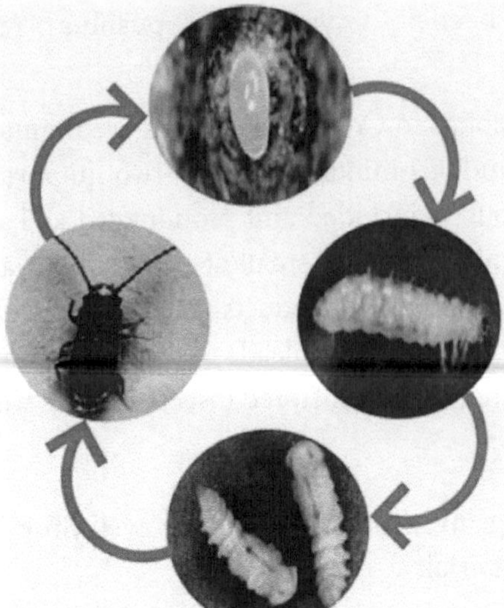

Fig. 1.13 Flat grain beetle: Cryptolestes pusillus (Schönherr) (Coleoptera : Laemophloeidae)

- Distribution and status: Worldwide present. But C. minutus is common in low temperature regions. C. pusillus confined to southern hemisphere.
- Host range: Almost all grain and grain products including oilseed cakes, dried fruits etc.
- Damaging stage: Both adults and grub. They are secondary pests and are unable to survive in sound uninjured grains.

Detection: Since the grubs a very minute, are unable to detect with naked eyes if less in number. However an adult are free moving in the grain and are rapidly multiplying, leads to trap easily with any probe traps.

e) Confused flour beetle: Tribolium confusum (Coleoptera :Tenebrionidae)

- Distribution and status: The confused flour beetle, originally of African origin, has a different distribution in that it occurs worldwide in cooler climates.

Egg

- Females lay an average of about 450 eggs, which are small and clear white.

Adult

- 1/8 to 1/4 inch in length, and reddish-brown in colour.
- The average life of adults is about one year.

Larvae

- Larvae (small brownish-white worms) hatch in five to twelve days.
- Full-grown in 30 to 40 days.

Pupae

- Pupal period is 5-8 days.

Host range: flour, cereals, meal, crackers, beans, spices, pasta, cake mix, dried pet food, dried flowers, chocolate, nuts, seeds.

Damaging stage: Both adults and larvae.

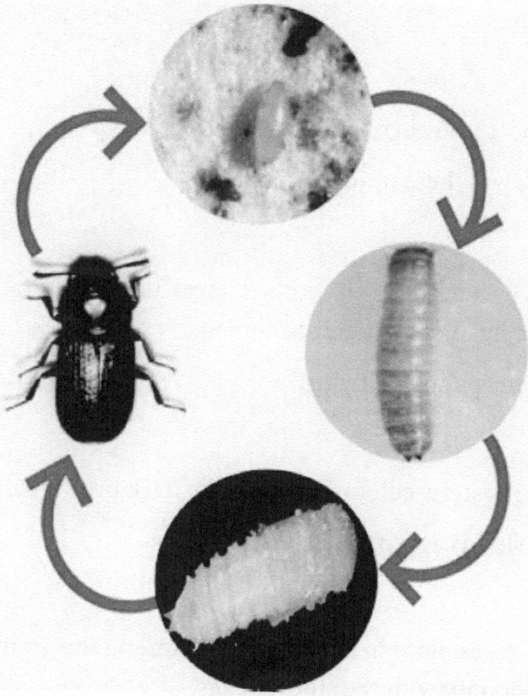

Fig.1.14 Confused flour beetle:Tribolium confusum (Coleoptera :Tenebrionidae)

SYMPTOM OF DAMAGE

Adults and larvae feed mainly on the germ of the cereal.

Detection: These insects are similar to rust-red flour beetle and are confusing. These insects are comparatively established well in lower temperatures than preferred by rust red flour beetle. The life cycle is almost similar for both of the species. Insects are generally prolonged with life cycle in winter and generally finishes total life cycle in 6 weeks under favorable conditions.

Dermestid Beetle

(a) Khapra beetle: Trogoderma granarium Everts (Coleoptera:Dermestidae)

These insects are having great quarantine importance worldwide and present in india. in many European countries.These insects are having zero tolerance pests. Insect can be able to attack any dried plant or animal matter but it prefers mainly grain and cereal products. In Indian conditions, khapra beetle is mainly suffering its severity in wheat storage but cab actually able to infect other commodities/grains also. Generally these insects breed from April to October in the active day. During warmer months i.e. July to October, the infestation is severe.

Egg

- Female laid 13-35 eggs, either in single or in cluster.
- Egg laying was there upto 1-7 days.

Adult

- Adult are reddish brown, convex, oval in shape practically no distinct division of head, thorax and abdomen.
- They are incap- able of flying.

Larvae

- The larvae is straw coloured having the dark brown hair band.
- Larval period is approximately 20- 40 days.

Pupae

- Pupation takes place in last larval skin among the grains or on the surface and overlapping with the storage bags.
- Pupal period is 4-6 days.

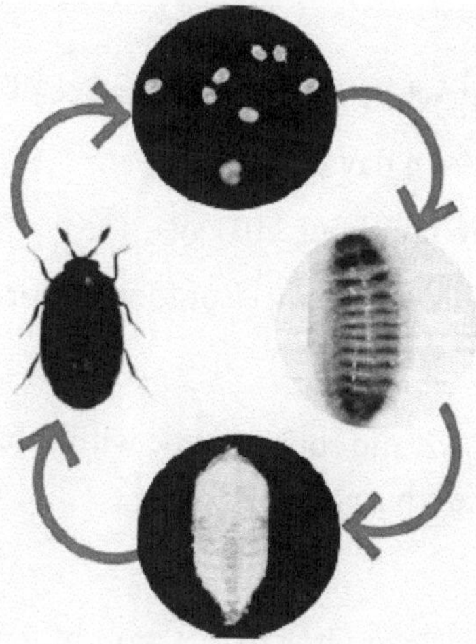

Fig. 1.15 Khapra beetle, Trogoderma granarium Everts (Coleoptera:Dermestidae)

- Distribution and status: Tropical and subtropical countries mainly found in hot and dry region. It prefers lower humidity and higher temperature.
- Host range: paddy, jowar, wheat, maize, oilseeds and pulses. Even wide range of stored products, packaged goods and handling facilities can also be affected.
- Damaging stage: Grubs are the voracious feeders. Adults are mainly involved in breeding. Detection:

Grubs are the damaging stage and feed on the germ portion by scratching even they finish entire grain. The exuviate was contaminating the grain maximum, with slight winnowing in handful of grains can give an idea about grub's presence. Except wheat, in other commodities grubs and adults can be easily differentiated. Since the grubs are positively thigmotactic, they can be collected by merely placing gunny bags on a heap of grains. Adults can be seen on the surface of grains while moving.

Miscellaneous beetles

(a) Cigarette beetle: Lasioderma serricorne (Coleoptera:Anobiidae)

Cigarette beetles are frequently found in warmer regions but having their distribution in warmer to colder regions but they are mainly in warmer regions. Though they are named as cigarette, they are not infested on stored tobacco leaves but also the large number of the other dry vegetable products, oilseeds, oilseed cakes, dried fruits etc.but practically all damage is done by larvae. Insect is active throughout the year in warm buildings, in temperate and subtropical regions and development slows down during winter as other insects.

Egg

- Female lays creamy white eggs on the surface of the stored material.
- The eggs hatch in 9-14 days.
- A single female can lay about 110 eggs.
- Eggs are laid in folds and crevices of food material.

Adult

- Adult are light brown and round beetle with thorax and head bent d o wnw ar d s , looking like a hump.

Larvae

- Newly hatched larvae are less than 1 mm in length and covered with fine hairs.
- Head of larvae is yellowish and has a semi-transparent and white body.
- Larvae damage the grains by making galleries.
- Larval period is approximately 25- 30 days.

Pupae

- The fully grown larvae transformed into pupa within a cell/cocoon made of chewed food material, newly formed pupa is glossy white.
- Pupal period is 4-6 days.

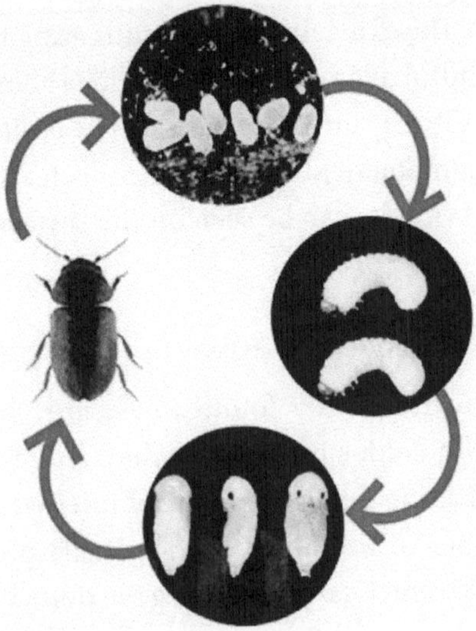

Fig. 1.16 Cigarette beetle, Lasioderma serricorne (Anobiidae: Coleoptera)

- Distribution and status: Cosmopolitan but prefers warm environment
- Host range: A wide range of commodities, but prefers to feed on cigarettes, chesools andchewing tobacco, chocolate, spices
- Damaging stage: Both grubs and adults Detection:

Presence of circular pinhead sized holes on processed tobacco in the typical symptom of attack. The grubs are tunneling inside the cigarettes and after transforming to adults they exit leaving hole on commodities.

(b) Drug store beetle: Stegobium paniceum (Linnaeus) (Coleoptera: Anobiidae)

Drugstore beetle, as name indicates they are mainly pest of stored spices. They have a tendency towards to feed on pharmacological/medicinal products as per ancient druggists but are feed on stored fruits, oilseeds, dried vegetables and plant products. These are primary pests of turmeric, ginger, pepper, coriander seeds etc.

- Distribution and status: Virtually cosmopolitan, more temperate than tropical
- Host range: A wide range of commodities, but prefers to feed on cigarettes, chesools and chewing tobacco, chocolate, spices.
- Damaging stage: Both grubs and adults Detection:

Infestation of drugstore beetles is similar to that of cigarette beetles.The exit hole of adult beetles appeared as shot like holes in infested products. These are also associated as an obligali symbiotic with fungus and are transferring from generations. The adults are active and can be seen flying or moving in the commodities. While, the grubs are feeding internally by making galleries.

Egg

- Adult female can lay upto 20 to 100 eggs. The eggs are laid on food material and are sometimes covered with yeast.
- Egg period is 8-10 days.

Adult

- Adults are small reddish brown beetles having wing. These are slightly larger than cigarette beetle. Adults are short lived.
- During after- noon to evening, these are active fliers. Adult female during favourable cond- itions can live upto 65 days.

Larvae

- The hatched out grubs are similar to cigarette beetles.
- The larva damages the commodity nearly by tunnel- ling.
- The larval period ranges from 4-20 weeks.
- In a warmer condition, they complete four generations per year.

Pupae

- Pupation is inside the damaged food products or in damaged products.
- The grub forms the cocoon and pupates inside.
- Pupal period is 4-6 days.

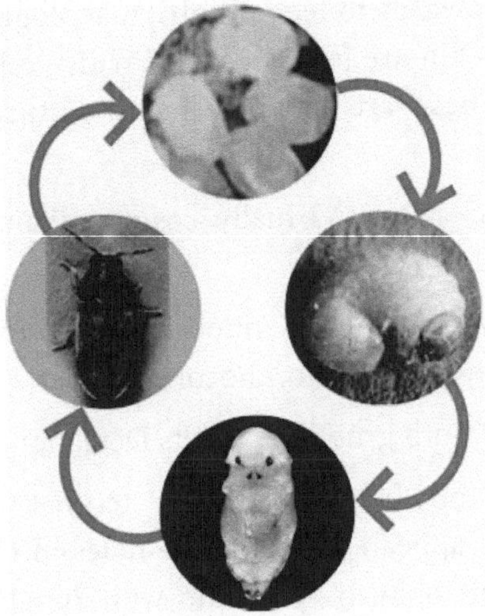

Fig. 1.17 Drug store beetle, Stegobium paniceum (Linnaeus) (Coleoptera: Anobiidae)

c) Groundnut bruchid: Caryedon serratus (Coleoptera: Chrysomelidae)

Bruchids are known to infest mainly groundnut commonly but are also reported on tamarind and four species of Fabaceae.These are prevalent in India under Gujarat, Karnataka, Andhra Pradesh,

Egg

- Small translucent milky white eggs laid on pods. Female can lay upto 60-70 eggs and the incubation period of eggs is 4 days.
- Irrespective of the egg density, only 2 adults emerge out of one-sided pod/ kernel and 4 to 5 adults from 2-3 seeded pods.

ADULT

- Adults are dark brown in color having sexual dimorphism.
- The posterior part of abdomen (Pygidium) is explored in female.
- Adult longevity is 15-20 days. Females are short lived than males.

Larvae

- Larvae upon hatch- ing started scraping pod/kernel surface and penetrate into pod to feed on kernels.
- The larvae having four instars and larval period are approximately 50 days.

Pupae

- The fully grown grub emerges out by making an exit hole and constructs a tough silken cocoon on the surface of pods/kernels.

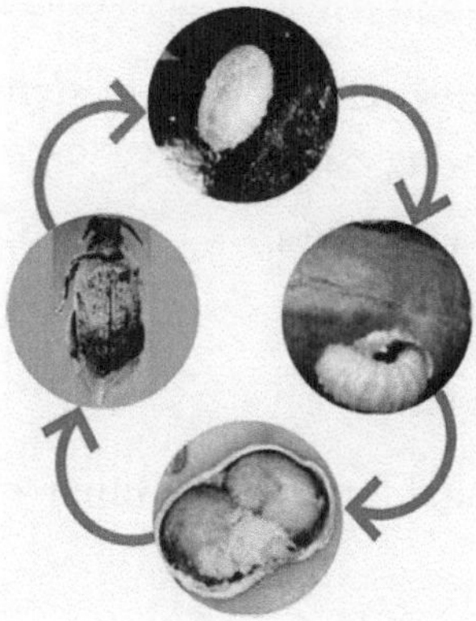

Fig. 1.18 Groundnut bruchid: Caryedon serratus (Coleoptera: Chrysomelidae)

Grubs pupate inside silken cocoon, having pupal period of 12 days.

Maharashtra and Tamil Nadu. Generally, infestation starts from field where the adults laying eggs on the pods and was carried to storage.

- Distribution and status: These are common in almost all groundnut growing countries. In India, it is most severe in Gujarat. Even its artificially tropicopolitan because of its association with Tamarindus indica.
- Host range: Groundnut, apart from it, also attacks four genera of Fabaceae. Tamarind is considered as most suitable host.
- Damaging stage: Mainly grubs Detection:

First sign of attack is appearance of windows cut into pod wall by larvae. The larvae burrows through pod wall and eat seeds. They damage both qualitatively and quantitatively. Increase in temperature because of higher metabolism rate. Indirectly also contaminate by aflatoxin spreading.

8. Grain moths

(a) Angoumois grain moth: Sitotroga cerealella (Olivier) (Lepidoptera: Gelechiidae)

The insect names come after its first notice during 1736 from Angoumois province of France. It is one of the most internal feeding destructor. Infestation starts from field itself during ripening/

Egg

- The female lays up to 150 eggs on the outside of kernels and in cracks.
- Egg period is 4-8 days.

Adult

- The adult is a buff, grey yellow, brown or straw coloured moth, measuring about 10-12 mm in wing expanse.
- Adult live for about 4 - 10 days.

Larvae

- Pupae

A full grown larva is about 5 mm long, with a white body and yellow brown head.

- Larvae capable of tunnelling sound grains.
- Larvae spins a silken cocoon inside the grain and changes into reddish brown pupa. Before pupation, larvae cut a circular opening on husk which is covered by silken cover to come out.

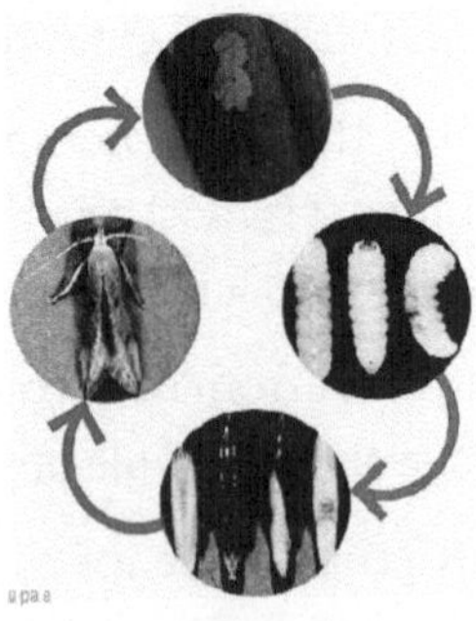

Fig. 1.19Angoumois grain moth:Sitotroga cerealella (Olivier) (Lepidoptera:Gelechiidae) [Ref. Chitrasrivastava, TNAU, USDA]

Pupal period is 9 -12 days

Milking stage. However, in storage infestation restricts to surface. Infestation produce abundant heat and moisture that may encourage mold growth and attract secondary pests.

- Distribution and status: Cosmopolitan, more abundant in warmer regions. However, in Indian conditions, pest is abundant in milder conditions.
- Host range: Regarded as most destructive internal feeder mainly attacking paddy, maize, sorghum etc. before harvest also.
- Damaging stage: Larvae, only whole cereals are attacked, greatest damage occurs in the upper layer grains in storage

Detection: Early infestation is difficult to detect because hole made by young insect is so small that it cannot be seen. Larva enters and its way in the grain, the turns about and spins a silken wets over the opening by which it entered that it is difficult to locate it. The appearance of moths in the stores and round holes in grains/heating of grains is first indication.

(b) Rice moth: Corcyra cephalonca (Stainton)(Lepidoptera: Pyralidae)

Rice moth is one of the key pests of rice, cocoa, confectionary, cereal grains etc. Adult insects do not feed and live for only 1 or 2 weeks.These are external feeders, which feed on grains by webbing them together. Rice moth is widely used for rearing of natural enemies in the laboratory use in field against crop pests as it is easier and cheaper to produce natural enemies on different stages of Corcyra trichogramma spp. is used for mass-breeding.

- Distribution and status: The rice moth is distributed in Asia, Africa, North America and Europe. In the larval stage, it is an important stored-grain pest in both India and Pakistan. Distributed well in all rice growing areas.
- Host range: Paddy, rice, wheat, maize, sorghum, millet, dried stored product.
- Damaging stage: Larva

Egg

- Eggs are whitish, oval in shape, 0.5 mm long and having an incubation period of 4-5 days.

Adult

- Adults light greyish-brown in colour, 12 mm long.

- Wing span of about 15 mm.

Pupae

- Pupal period is about 10 days but may extend to 40-50 days to tide over winter.

Larvae

- Full grown larva is pale whitish in colour
- 15 mm long with short scattered hairs and no markings on body.

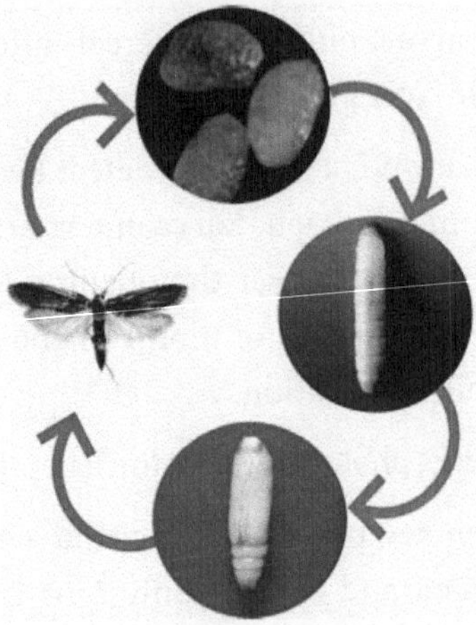

Fig. 1.20 Rice moth: Corcyra cephalonca (Stainton) (Lepidoptera: Pyralidae)

Detection: Larvae cause the damage by webbing together grains and forming lump and feed from inside it. Larvae leave a lot of webbing in the grains before pupation, causing excessive lumping, which reduces marketing quality of the grains.C. cephalonica adults can be seen resting on store surfaces and have a peak of flight activity at dusk. The larvae crawl over the stored food and in the last instar construct cocoons that may be found within the stored food, on sack surfaces and store structures. Infestations cause a bad smell.

9. Indian meal moth: Plodia interpunctella (Hubner) (Lepidoptera: Pyralidae)

Indian meal moth commonly called as weevil moth, pantry moth or flour moth. It is generally confused with almond moth (Candracantella). The larvae are commonly known as waxworms.

Egg

- Eggs are laid singly or in clusters and stick to various foods by a sticky secretion.

- Eggs hatch in 4-20 days.

Adult

- Adults mate immediately after emergence and start the life cycle.
- Female moth can lay 30-350 eggs in a minute. Adults generally do not feed.

Larvae

- Larvae having 5-7 instars. Newly hatched larvae feed on grain while mature larvae feed on grain germ.
- White colour larvae having polychromy in pink, brown and greenish.
- Fully grown larvae able to spin webs and leave silk threads in their path of travel.

Pupae

- The fully grown larvae make threaded cocoons and pupate inside. Sometimes pupa is often seen on grain surface and on wall of bins.
- Pupal period is 4-10 days

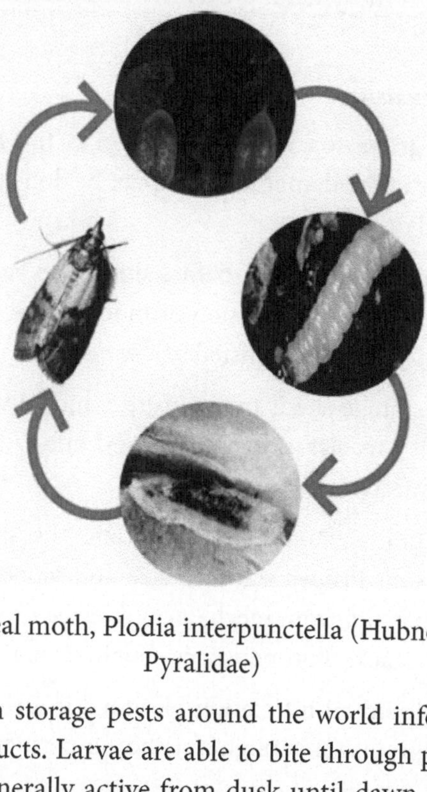

Fig.1.21 Indian meal moth, Plodia interpunctella (Hubner) (Lepidoptera: Pyralidae)

These are common storage pests around the world infecting stored cereals, fruits and similar products. Larvae are able to bite through plastic and cardboard also. The moths are generally active from dusk until dawn, rests during day.The adults are easily distinguished from other grain pests with peculiar markings of the forewings which are reddish brown with a copper luster on body ends.

- Distribution and status: A native to America, but now a cosmopolitan species. Adapted to wide range of climates around the world.
- Host range: Wider host range includes stored grains and pulses, dried fruits and nuts, dried vegetables and processed foods.
- Damaging stage: Larva

Detection: Larvae creates more severe problem by feeding and contaminating produce through faeces. Larvae feed by webbing. The webbing may also cause condensation which leads to damaging molds. Crawling caterpillars completely web over the surface of heap of grains with silken threads. The adults fly from one bin to another and spread the infestation.

General management of stored product insects

Although, insect management options like physical, mechanical, biological and chemical practices are available, in India fumigation was practiced since long time in storage. Some of the general preventive and curative management options are listed below.

I. Preventive measures

- Drying of the grains to a moisture range of 10 to 14% based on grain type, avoid majority of the damage. Drying can be done under sunlight or using any developed dryers.
- After harvesting of the grains before shifting to storage, certain measures have to take the initial infestation or inoculum of storages pests. Some of the preventive measures are listed below,
- Sanitation of storage site is prerequisite, which involves removing of dirt, debris, foreign particles, other insects and infested grains that will reduce avoid initial infestation.
- Handling of grains should be proper and the storage structure should not be damaged. Proper stacking size and wooden dunnage should be maintained to avoid from mechanical damage or to keep away from wall. Between each stacks a proper space should be maintained.
- Sometimes the initial infestation observed in field can be managed with exposing the grains for sunlight for shorter period.

- Using of improved storage structures. In Indian conditions grains are storing variedly from small metal bin to bulk storage in gunny bags under warehouses. Hermetic storage concept developing can be exploited for the future. Jute bags are generally using, which are additionally suggesting to amalgamate with polythene lining.
- Newer lots have to store separately and should have isolated distance between old stocks.
- Several improved bins are available, which are developed by different institutes for Indian conditions like, Pusa bin, PAU bin, TNAU bin, etc. can be used.
- Disinfestation of bulk and bag storage structures using insecticides is also an important practice to be done prior to storage and a layer of insecticide spray can be suggested

immediately after storage. Since, dichlorvos was banned using in warehouses, presently deltamethrin and malathion are the available options for disinfectant and for surface sprays.

II. Curative measures

a. *Physical methods*

1. Temperature management: Increasing or decreasing the temperature can alter the insect growth and metabolism. Optimal temperature for most of the storage insects is between 25 and 33° C. deviations from these points will slower down the growth and continuously to death. Passing of dry or wet heat in the storage system or refrigerated aeration yield better results. Even it can be achieved using high frequency waves.
2. Mixing of inert dusts: Inert dusts like clays, sand, ash, minerals, silica (silicon dioxide) are effective in managing the insects by moisture loss of the insect body by abrasion.Dusts that contain natural silica, such as diatomaceous earth (DE), are commercially available and using in many developing countries either to manage or to improve fumigation efficiency. Activated clay (kaolin) has also been used in protecting grains from the attack of storage insects.
3. Irradiation: Radiations in lower dose can able to kill or sterilizes the common grain pests, and even the eggs deposited inside the grains. Radiations like microwaves, x- rays, etc. are utilising in several forms to treat the grains before storage to disinfest them.

4. Use of controlled atmosphere: A novel methodology where the storage atmosphere generally contains 78% Nitrogen (N2), 21% Oxygen (O2) and 0.03% carbon dioxide (CO2). This proportion can be altered mechanically by altered atmosphere which interfere with the normal insect metabolism which achieve mortality.
5. Mechanical devices, traps, etc.: Devices developed for monitoring and mass trapping such as entoleters, which are using in flour mills. TNAU under AICRP on Post-Harvest Engineering &Technology has developed different types of insects traps (TNAU stored insect kit). Apart from these some other traps like Probe trap, Pulse Beetle Trap, Light traps, Sticky traps, Bait traps and Pheromone were also developed and tested in some of the countries.

b..... Biological control:

1. Semiochemicals: Much of the behavior of insect pests of stored products is associated with the search for food, sexual partners and egg-laying sites, together with defensive activities to protect them from adverse environmental conditions and natural enemies. This behavior is modified by chemical signals produced either by the insects themselves or by other organisms including the plants upon which they feed. This includes pheromones and allelochemicals. Insect traps are available for some of the insects like flour beetles and lepidopteran insects.
2. Botanicals: Plant extracts or products having insecticidal properties like Neem leaf powder, black pepper, turmeric powder, Sweet Flag Rhizome powder etc were also proved their effectiveness in amangingthe stored grain pests. General dosage is 10g / kg of grains should be mixed. Different formulations like tablets, pellets, oils, etc are developing and are required commercialisation.
3. Biopesticides: Some of the commercially available entomopathogenic fungi against field crop pests like Beauveriabassiana, Metarhiziumanisopliae and bacterium - Bacillus thuringiensis (Bt) were tested majorly against stored-grain pests especially beetles.

2

PEST MANAGEMENT

A pest is "a plant or animal detrimental to humans or human concerns (as agriculture or livestock production)"; alternative meanings include organisms that cause nuisance and epidemic disease associated with high mortality (specifically: plague). In its broadest sense, a pest is a competitor of humanity.

A pest is any living organism, whether animal, plant or fungus, which is invasive or troublesome to plants or animals, human or human concerns, livestock, or human structures. It is a loose concept, as an organism can be a pest in one setting but beneficial, domesticated or acceptable in another.

Animals are called pests when they cause damage to agriculture by feeding on crops or parasitisinglivestock, such as codling moth on apples, or boll weevil on cotton. An animal could also be a pest when it causes damage to a wild ecosystem or carries germs within human habitats. Examples of these include those organisms which vector human disease, such as rats and fleas which carry the plaguedisease, mosquitoes which vector malaria, and ticks which carry Lyme disease.

A species can be a pest in one setting but beneficial or domesticated in another (for example, European rabbits introduced to Australia caused ecological damage beyond the scale they inflicted in their natural habitat). Many weeds are also seen as useful under certain conditions, for instance Patterson's curse is often valued as food for honeybees and as a wildflower, even though it can poison livestock.

The animal groups of greatest importance as pests (in order of economic importance) are insects, mites, nematodes and gastropods.

The term "plant pest" has a specific definition in terms of the International Plant Protection Convention and phytosanitary measures worldwide. A pest is any species, strain or biotype of plant, animal, or pathogenic agent injurious to plants or plant products. Plants may be considered pests themselves if an invasive species

Pest control is the regulation or management of a species defined as a pest, a member of the animal kingdom that impacts adversely on human activities. The human response depends on the importance of the damage done, and will range from tolerance, through deterrence and management, to attempts to completely eradicate the pest. Pest control measures may be performed as part of an integrated pest management strategy.

In agriculture, pests are kept at bay by cultural, chemical and biological means. Ploughing and cultivation of the soil before sowing reduces the pest burden and there is a modern trend to limit the use of pesticides as far as possible. This can be achieved by monitoring the crop, only applying insecticides when necessary, and by growing varieties and crops which are resistant to pests. Where possible, biological means are used, encouraging the natural enemies of the pests and introducing suitable predators or parasites.

In homes and urban environments, the pests are the rodents, birds, insects and other organisms that share the habitat with humans, and that feed on and spoil possessions. Control of these pests is attempted through exclusion, repulsion, physical removal or chemical means. Alternatively, various methods of biological control can be used including sterilisation programmes.

Pest control is at least as old as agriculture, as there has always been a need to keep crops free from pests. As long ago as 3000 BC in Egypt, cats were used to control pests of grain stores such as rodents. Ferrets were domesticated by 500 AD in Europe for use as mousers. Mongooses were introduced into homes to control rodents and snakes, probably by the ancient Egyptians. The conventional approach was probably the first to be employed, since it is comparatively easy to destroy weeds by burning them or ploughing them under, and to kill larger competing herbivores. Techniques such as crop rotation, companion planting (also known as intercropping or mixed cropping), and the selective breeding of pest-resistant cultivars have a long history.

Red weaver ants, here feeding on a snail, have been used to control pests in China, Southeast Asia and Africa for many centuries.

Chemical pesticides were first used around 2500 BC, when the Sumerians used sulphur compounds as insecticides. Modern pest control was stimulated by the spread across the United States of the Colorado potato beetle. After much discussion, arsenicalcompounds were used to control the beetle and the predicted poisoning of the human population did not occur. This led the way to a widespread acceptance of insecticides across the continent. With the industrialisation and mechanization of agriculture in the 18th and 19th centuries, and the introduction of the insecticides pyrethrum and derris, chemical pest control became widespread. In the 20th century, the discovery of several synthetic insecticides, such as DDT, and herbicides boosted this development.

Biological control is first recorded around 300 AD in China, when colonies of weaver ants, Oecophylla smaragdina, were intentionally placed in citrus plantations to control beetles and caterpillars. Also in China, ducks were used in paddy fields to consume pests, as illustrated in ancient cave art. In 1762, an Indian mynah was brought to Mauritius to control locusts, and about the same time, citrus trees in Burma were connected by bamboos to allow ants to pass between them and help control caterpillars. In the 1880s, ladybirds were used in citrus plantations in California to control scale insects, and other biological control experiments followed. The introduction of DDT, a cheap and effective compound, put an effective stop to biological control experiments. By the 1960s, problems of resistance to chemicals and damage to the environment began to emerge, and biological control had a renaissance. Chemical pest control is still the predominant type of pest control today, although a renewed interest in traditional and biological pest control developed towards the end of the 20th century and continues to this da

Biological control is a method of controlling pests such as insects, mites, weeds and plant diseases using other organisms. It relies on predation, parasitism, herbivory, or other natural mechanisms, but typically also involves an active human management role. It can be an important component of integrated pest management (IPM) programs.

There are three basic types of biological pest control strategies: importation (sometimes called classical biological control), in which a natural enemy of a pest is introduced in the hope of achieving control; augmentation, in which locally-occurring natural enemies are bred and released to improve control; and conservation, in which measures are taken to increase natural enemies, such as by planting nectar-producing crop plants in the borders of rice fields.

Natural enemies of insect pests, also known as biological control agents, include predators, parasitoids, pathogens, and competitors. Biological control agents of plant diseases are most often referred to as antagonists. Biological control agents of weeds include seed predators, herbivoresand plant pathogens.

Biological control can have side-effects on biodiversity through attacks on non-target species by any of the same mechanisms, especially when a species is introduced without thorough understanding of the possible consequences.

The term "biological control" was first used by Harry Scott Smith at the 1919 meeting of the Pacific Slope Branch of the American Association of Economic Entomologists, in Riverside, California. It was brought into more widespread use by the entomologist Paul H. DeBach (1914–1993) who worked on citrus crop pests throughout his life. However, the practice has previously been used for centuries. The first report of the use of an insect species to control an insect pest comes from "Nan Fang Cao Mu Zhuang" (WS¹eIƒ(gÀr Plants of the Southern Regions) (ca. 304 AD), attributed to Western Jin dynasty botanist Ji Han (G]+T, 263–307), in which it is mentioned that "Jiaozhi people sell ants and their nests attached to twigs looking like thin cotton envelopes, the reddish-yellow ant being larger than normal. Without such ants, southern citrus fruits will be severely insect-damaged". The ants used are known as huang gan (huang = yellow, gan = citrus) ants (Oecophylla smaragdina). The practice was later reported by Ling Biao Lu Yi (late Tang Dynasty or Early Five Dynasties), in Ji Le Pian by Zhuang Jisu (Southern Song Dynasty), in the Book of Tree Planting by Yu Zhen Mu (Ming Dynasty), in the book Guangdong Xing Yu (17th century), Lingnan by Wu Zhen Fang (Qing Dynasty), in Nanyue Miscellanies by Li Diao Yuan, and others.

Biological control techniques as we know them today started to emerge in the 1870s. During this decade, in the USA, the Missouri State Entomologist C. V. Riley and the Illinois State Entomologist W. LeBaron began within-state redistribution of parasitoids to control crop pests. The first international shipment of an insect as biological control agent was made by Charles V. Riley in 1873, shipping to France the predatory mites Tyroglyphus phylloxera to help fight the grapevine phylloxera (Daktulosphaira vitifoliae) that was destroying grapevines in France. The United States Department of Agriculture (USDA) initiated research in classical biological control following the establishment of the Division of Entomology in 1881, with C. V. Riley as Chief. The first importation of a parasitoidal wasp into the United States was that of the braconid Cotesia glomerata in 1883–1884, imported from Europe to control the invasive cabbage white butterfly, Pieris rapae. In 1888–1889 the vedalia beetle, Rodolia cardinalis, a lady beetle, was introduced from Australia to California to control the cottony cushion scale, Icerya purchasi. This had become a major problem for the newly developed citrus industry in California, but by the end of 1889 the cottony cushion scale population had already declined. This great success led to further introductions of beneficial insects into the USA.

In 1905 the USDA initiated its first large-scale biological control program, sending entomologists to Europe and Japan to look for natural enemies of the gypsy moth, Lymantria dispar dispar, and brown-tail moth, Euproctis chrysorrhoea, invasive pests of trees and shrubs. As a result, nine parasitoids (solitary wasps) of gypsy moth, seven of brown-tail moth, and two predators of both moths became established in the USA. Although the gypsy moth was not fully controlled by these natural enemies, the frequency, duration, and severity of its outbreaks were reduced and the program was regarded as successful. This program also led to the development of many concepts, principles, and procedures for the implementation of biological control programs.

Prickly pear cacti were introduced into Queensland, Australia as ornamental plants, starting in 1788. They quickly spread to cover over 25 million hectares of Australia by 1920, increasing by 1 million hectares per year. Digging, burning and crushing all proved ineffective. By 1914, two control agents were introduced to help control the spread of the plant, the cactus moth Cactoblastis cactorum, and the scale insect Dactylopius. By 1917, most areas of prickly pear had been destroyed.

The first reported case of a classical biological control attempt in Canada involves the parasitoidal wasp Trichogramma minutum. Individuals were caught in New York State and released in Ontario gardens in 1882 by William Saunders, trained chemist and first Director of the Dominion Experimental Farms, for controlling the invasive currantworm Nematus ribesii. Between 1884 and 1908, the first Dominion Entomologist, James Fletcher, continued introductions of other parasitoids and pathogens for the control of pests in Canada.

TYPES OF BIOLOGICAL PEST CONTROL

There are three basic biological pest control strategies: importation (classical biological control), augmentation and conservation.

Importation

Importation or classical biological control involves the introduction of a pest's natural enemies to a new locale where they do not occur naturally. Early instances were often unofficial and not based on research, and some introduced species became serious pests themselves. To be most effective at controlling a pest, a biological control agent requires a colonizing ability which allows it to keep pace with changes to the habitat in space and time. Control is greatest if the agent has temporal persistence, so that it can maintain its population even in the temporary absence of the target species, and if it is an opportunistic forager, enabling it to rapidly exploit a pest population.

Joseph Needham noted a Chinese text dating from 304 AD, Records of the Plants and Trees of the Southern Regions, by Hsi Han, which describes mandarin oranges protected by large reddish-yellow citrus ants which attack and kill insect pests of the orange trees. The citrus ant (Oecophylla smaragdina) was rediscovered in the 20th century, and since 1958 has been used in China to protect orange groves.

One of the earliest successes in the west was in controlling Icerya purchasi (cottony cushion scale) in Australia, using a predatory insect Rodolia cardinalis (the vedalia beetle). This success was repeated in California using the beetle and a parasitoidal fly, Cryptochaetum iceryae.

Damage from Hypera postica, the alfalfa weevil, a serious introduced pest of forage, was substantially reduced by the introduction of natural enemies. 20 years after their introduction the population of weevils in the alfalfa area treated for alfalfa weevil in the Northeastern United Statesremained 75 percent down.

Alligator weed was introduced to the United States from South America. It takes root in shallow water, interfering with navigation, irrigation, and flood control. The alligator weed flea beetle and two other biological controls were released in Florida, greatly reducing the amount of land covered by the plant.Another aquatic weed, the giant salvinia (Salvinia molesta) is a serious pest, covering waterways, reducing water flow and harming native species. Control with the salvinia weevil (Cyrtobagous salviniae) is effective in warm climates, and in Zimbabwe, a 99% control of the weed was obtained over a two-year period.

Small commercially reared parasitoidal wasps, Trichogramma ostriniae, provide limited and erratic control of the European corn borer (Ostrinia nubilalis), a serious pest. Careful formulations of the bacterium Bacillus thuringiensis are more effective.

The population of Levuana iridescens, the Levuana moth, a serious coconut pest in Fiji, was brought under control by a classical biological control program in the 1920s.

Augmentation

Augmentation involves the supplemental release of natural enemies that occur in a particular area, boosting the naturally occurring populations there. In inoculative release, small numbers of the control agents are released at intervals to allow them to reproduce, in the hope of setting up longer-term control, and thus keeping the pest down to a low level, constituting prevention rather than cure. In inundative release, in contrast, large numbers are released in the hope of rapidly reducing a damaging pest population, correcting a problem that has already arisen. Augmentation can be effective, but is not guaranteed to work, and depends on the precise details of the interactions between each pest and control agent.

An example of inoculative release occurs in the horticultural production of several crops in greenhouses. Periodic releases of the parasitoidal wasp, Encarsia formosa, are used to control greenhouse whitefly, while the predatory mite Phytoseiulus persimilis is used for control of the two-spotted spider mite.

The egg parasite Trichogramma is frequently released inundatively to control harmful moths. Similarly, Bacillus thuringiensis and other microbial insecticides are used in large enough quantities for a rapid effect. Recommended release rates for Trichogramma in vegetable or field crops range from 5,000 to 200,000 per acre (1 to 50 per square metre) per week according to the level of pest infestation. Similarly, nematodes that kill insects (that are entomopathogenic) are released at rates of millions and even billions per acre for control of certain soil-dwelling insect pests.

Conservation

The conservation of existing natural enemies in an environment is the third method of biological pest control. Natural enemies are already adapted to the habitat and to the target pest, and their conservation can be simple and cost-effective, as when nectar-producing crop plants are grown in the borders of rice fields. These provide nectar to support parasitoids and predators of planthopper pests and have been demonstrated to be so effective (reducing pest densities by 10- or even 100-fold) that farmers sprayed 70% less insecticides and enjoyed yields boosted by 5%. Predators of aphids were similarly found to be present in tussock grasses by field boundary hedges in England, but they spread too slowly to reach the centres of fields. Control was improved by planting a metre-wide strip of tussock grasses in field centres, enabling aphid predators to overwinter there.

Fig 2.1

An inverted flowerpot filled with straw to attract earwigs

Cropping systems can be modified to favor natural enemies, a practice sometimes referred to as habitat manipulation. Providing a suitable habitat, such as a shelterbelt, hedgerow, or beetle bank where beneficial insects such as parasitoidal wasps can live and reproduce, can help ensure the survival of populations of natural enemies. Things as simple as leaving a layer of fallen leaves or mulch in place provides a suitable food source for worms and provides a shelter for insects, in turn being a food source for such beneficial mammals as hedgehogs and shrews. Compost piles and stacks of wood can provide shelter for invertebrates and small mammals. Long grass and ponds support amphibians. Not removing dead annuals and non-hardy plants in the autumn allows insects to make use of their hollow stems during winter. In California, prune trees are sometimes planted in grape vineyards to provide an improved overwintering habitat or refuge for a key grape pest parasitoid. The providing of artificial shelters in the form of wooden caskets, boxes or flowerpots is also sometimes undertaken, particularly in gardens, to make a cropped area more attractive to natural enemies. For example, earwigs are natural predators which can be encouraged in gardens by hanging upside-down flowerpots filled with straw or wood wool. Green lacewings can be encouraged by using plastic bottles with an open bottom and a roll of cardboard inside. Birdhouses enable insectivorous birds to nest; the most useful birds can be attracted by choosing an opening just large enough for the desired species.

In cotton production, the replacement of broad-spectrum insecticides with selective control measures such as Bt cotton can create a more favorable environment for natural enemies of cotton pests due to reduced insecticide exposure risk. Such predators or parasitoids can control pests not affected by the Bt protein. Reduced prey quality and abundance associated increased control from Bt cotton can also indirectly decrease natural enemy populations in some cases, but the percentage of pests eaten or parasitized in Bt and non-Bt cotton are often similar.

Predators

Predators are mainly free-living species that directly consume a large number of prey during their whole lifetime. Given that many major crop pests are insects, many of the predators used in biological control are insectivorous species. Lady beetles, and in particular their larvae which are active between May and July in the northern hemisphere, are voracious predators of aphids, and also consume mites, scale insects and small caterpillars. The spotted lady beetle (Coleomegilla maculata) is also able to feed on the eggs and larvae of the Colorado potato beetle (Leptinotarsa decemlineata).

The larvae of many hoverfly species principally feed upon aphids, one larva devouring up to 400 in its lifetime. Their effectiveness in commercial crops has not been studied.

Several species of entomopathogenic nematode are important predators of insect and other invertebrate pests.Phasmarhabditis hermaphrodita is a microscopic nematodethat kills slugs. Its complex life cycle includes a free-living, infective stage in the soil where it becomes associated with a pathogenic bacteria such as Moraxella osloensis. The nematode enters the slug through the posterior mantle region, thereafter feeding and reproducing inside, but it is the bacteria that kill the slug. The nematode is available commercially in Europe and is applied by watering onto moist soil.

Species used to control spider mites include the predatory mites Phytoseiulus persimilis, Neoseilus californicus, and Amblyseius cucumeris, the predatory midge Feltiella acarisuga, and a ladybird Stethorus punctillum. The bug Orius insidiosus has been successfully used against the two-spotted spider mite and the western flower thrips (Frankliniella occidentalis).

PARASITOIDS

Parasitoids lay their eggs on or in the body of an insect host, which is then used as a food for developing larvae. The host is ultimately killed. Most insect parasitoids are wasps or flies, and many have a very narrow host range. The most important groups are the ichneumonid wasps, which mainly use caterpillars as hosts; braconid wasps, which attack caterpillars and a wide range of other insects including aphids; chalcid wasps, which parasitize eggs and larvae of many insect species; and tachinid flies, which parasitize a wide range of insects including caterpillars, beetle adults and larvae, and true bugs.

Encarsia formosa is a small predatory chalcid wasp which is a parasitoid of whitefly, a sap-feeding insect which can cause wilting and black sooty moulds in glasshouse vegetable and ornamental crops. It is most effective when dealing with low level infestations, giving protection over a long period of time. The wasp lays its eggs in young whitefly 'scales', turning them black as the parasite larvae pupate.Gonatocerus ashmeadi (Hymenoptera: Mymaridae) has been introduced to control the glassy-winged sharpshooter Homalodisca vitripennis (Hemiptera: Cicadellidae) in French Polynesia and has successfully controlled ~95% of the pest density.

Parasitoids are among the most widely used biological control agents. Commercially, there are two types of rearing systems: short-term daily output with high production of parasitoids per day, and long-term, low daily output systems. In most instances, production will need to be matched with the appropriate release dates when susceptible host species at a suitable phase of development will be available. Larger production facilities produce on a yearlong basis, whereas some facilities produce only seasonally. Rearing facilities are usually a significant distance from where the agents are to be used in the field, and transporting the parasitoids from the point of production to the point of use can pose problems. Shipping conditions can be too hot, and even vibrations from planes or trucks can adversely affect parasitoids.

Pathogens

Pathogenic micro-organisms include bacteria, fungi, and viruses. They kill or debilitate their host and are relatively host-specific. Various microbial insect diseases occur naturally, but may also be used as biological pesticides. When naturally occurring, these outbreaks are density-dependent in that they generally only occur as insect populations become denser.

Bacteria

Bacteria used for biological control infect insects via their digestive tracts, so they offer only limited options for controlling insects with sucking mouth parts such as aphids and scale insects. Bacillus thuringiensis is the most widely applied species of bacteria used for biological control, with at least four sub-species used against Lepidopteran (moth, butterfly), Coleopteran (beetle) and Dipteran (true fly) insect pests. The bacterium is available to organic farmers in sachets of dried spores which are mixed with water and sprayed onto vulnerable plants such as brassicas and fruit trees. Genes from B. thuringiensis have also been incorporated into transgenic crops, making the plants express some of the bacterium's toxins, which are proteins. These confer resistance to insect pests and thus reduce the necessity for pesticide use. If pests develop resistance to the toxins in these crops, B. thuringiensis will become useless in organic farming also. The bacterium Paenibacillus popilliae which causes milky spore disease has been found useful in the control of Japanese beetle, killing the larvae. It is very specific to its host species and is harmless to vertebrates and other invertebrates.

FUNGI

Entomopathogenic fungi, which cause disease in insects, include at least 14 species that attack aphids. Beauveria bassiana is mass-produced and used to manage a wide variety of insect pests including whiteflies, thrips, aphids and weevils. Lecanicillium spp. are deployed against white flies, thrips and aphids.

Metarhizium spp. are used against pests including beetles, locusts and other grasshoppers, Hemiptera, and spider mites. Paecilomyces fumosoroseus is effective against white flies, thrips and aphids; Purpureocillium lilacinus is used against root-knot nematodes, and 89 Trichodermaspecies against certain plant pathogens. Trichoderma viride has been used against Dutch elm disease, and has shown some effect in suppressing silver leaf, a disease of stone fruits caused by the pathogenic fungus Chondrostereum purpureum.

The fungi Cordyceps and Metacordyceps are deployed against a wide spectrum of arthropods.Entomophaga is effective against pests such as the green peach aphid.

Several members of Chytridiomycota and Blastocladiomycota have been explored as agents of biological control. From Chytridiomycota, Synchytrium solstitiale is being considered as a control agent of the yellow star thistle (Centaurea solstitialis) in the United States.

Viruses

Baculoviruses are specific to individual insect host species and have been shown to be useful in biological pest control. For example, the Lymantria dispar multicapsid nuclear polyhedrosis virus has been used to spray large areas of forest in North America where larvae of the gypsy moth are causing serious defoliation. The moth larvae are killed by the virus they have eaten and die, the disintegrating cadavers leaving virus particles on the foliage to infect other larvae.

A mammalian virus, the rabbit haemorrhagic disease virus was introduced to Australia to attempt to control the European rabbit populations there. It escaped from quarantine and spread across the country, killing large numbers of rabbits. Very young animals survived, passing immunity to their offspring in due course and eventually producing a virus-resistant population. Introduction into New Zealand in the 1990s was similarly successful at first, but a decade later, immunity had developed and populations had returned to pre-RHD levels.

Algae

Lagenidium giganteum is a water-borne mould that parasitizes the larval stage of mosquitoes. When applied to water, the motile spores avoid unsuitable host species and search out suitable mosquito larval hosts. This alga has the advantages of a dormant phase, resistant to desiccation, with slow-release characteristics over several years. Unfortunately, it is susceptible to many chemicals used in mosquito abatement programmes.

Competitors

The legume vine Mucuna pruriens is used in the countries of Benin and Vietnam as a biological control for problematic Imperata cylindricagrass: the vine is extremely vigorous and suppresses neighbouring plants by out-competing them for space and light. Mucuna pruriens is said not to be invasive outside its cultivated area. Desmodium uncinatum can be used in push-pull farming to stop the parasitic plant, witchweed (Striga).

The Australian bush fly, Musca vetustissima, is a major nuisance pest in Australia, but native decomposers found in Australia are not adapted to feeding on cow dung, which is where bush flies breed. Therefore, the Australian Dung Beetle Project (1965–1985), led by George Bornemisszaof the Commonwealth Scientific and Industrial Research Organisation, released forty-nine species of dung beetle, to reduce the amount of dung and therefore also the potential breeding sites of the fly.

COMBINED USE OF PARASITOIDS AND PATHOGENS

In cases of massive and severe infection of invasive pests, techniques of pest control are often used in combination. An example is the emerald ash borer, Agrilus planipennis, an invasive beetle from China, which has destroyed tens of millions of ash trees in its introduced range in North America. As part of the campaign against it, from 2003 American scientists and the Chinese Academy of Forestry searched for its natural enemies in the wild, leading to the discovery of several parasitoid wasps, namely Tetrastichus planipennisi, a gregarious larval endoparasitoid, Oobius agrili, a solitary, parthenogenic egg parasitoid, and Spathius agrili, a gregarious larval ectoparasitoid. These have been introduced and released into the United States of America as a possible biological control of the emerald ash borer. Initial results for Tetrastichus planipennisihave shown promise, and it is now being released along with Beauveria bassiana, a fungal pathogen with known insecticidal properties.

Difficulties

Many of the most important pests are exotic, invasive species that severely impact agriculture, horticulture, forestry and urban environments. They tend to arrive without their co-evolved parasites, pathogens and predators, and by escaping from these, populations may soar. Importing the natural enemies of these pests may seem a logical move but this may have unintended consequences; regulations may be ineffective and there may be unanticipated effects on biodiversity, and the adoption of the techniques may prove challenging because of a lack of knowledge among farmers and growers.

SIDE-EFFECTS

Biological control can affect biodiversity through predation, parasitism, pathogenicity, competition, or other attacks on non-target species.An introduced control does not always target only the intended pest species; it can also target native species. In Hawaii during the 1940s parasitic wasps were introduced to control a lepidopteran pest and the wasps are still found there today. This may have a negative impact on the native ecosystem, however, host range and impacts need to be studied before declaring their impact on the environment.

Vertebrate animals tend to be generalist feeders, and seldom make good biological control agents; many of the classic cases of "biocontrol gone awry" involve vertebrates. For example, the cane toad(Rhinella marina) was intentionally introduced to Australia to control the greyback cane beetle(Dermolepida albohirtum), and other pests of sugar cane. 102 toads were obtained from Hawaii and bred in captivity to increase their numbers until they were released into the sugar cane fields of the tropic north in 1935. It was later discovered that the toads could not jump very high and so were unable to eat the cane beetles which stayed up on the upper stalks of the cane plants. However the toad thrived by feeding on other insects and it soon spread very rapidly; it took over native amphibian habitatand brought foreign disease to native toads and frogs, dramatically reducing their populations. Also when it is threatened or handled, the cane toad releases poison from parotoid glands on its shoulders; native Australian species such as goannas, tiger snakes, dingos and northern quolls that attempted to eat the toad were harmed or killed. However, there has been some recent evidence that native predators are adapting, both physiologically and through changing their behaviour, so in the long run, their populations may recover.

Rhinocyllus conicus, a seed-feeding weevil, was introduced to North America to control exotic musk thistle (Carduus nutans) and Canadian thistle (Cirsium arvense). However the weevil also attacks native thistles, harming such species as the endemic Platte thistle (Cirsium neomexicanum) by selecting larger plants (which reduced the gene pool), reducing seed production and ultimately threatening the species' survival.

The small Asian mongoose (Herpestus javanicus) was introduced to Hawaii in order to control the rat population. However the mongoose was diurnal, and the rats emerged at night; so it preyed on the endemic birds of Hawaii, especially their eggs, more often than it ate the rats, and now both rats and mongooses threaten the birds. This introduction was undertaken without understanding the consequences of such an action. No regulations existed at the time, and more careful evaluation should prevent such releases now.

The sturdy and prolific eastern mosquitofish (Gambusia holbrooki) is a native of the southeastern United States and was introduced around the world in the 1930s and 40s to feed on mosquito larvae and thus combat malaria. However, it has thrived at the expense of local species, causing a decline of endemic fish and frogs through competition for food resources, as well as through eating their eggs and larvae. In Australia, the mosquitofish is the subject of discussion as to how best to control it; in 1989 it was said that "biological population control is well beyond present capabilities", and this remains the position.

Humans have always been in direct competition with a myriad of insects, pests from our ancestral beginning. This competition for food with insects intensified when humans began to cultivate plants converting the natural ecosystem to an agroecosystem. Also insects serve as vectors of various diseases caused by bacterial, filarial nematode, protozoans and viruses. Therefore control of insects posed a major concern for the development of the economy. In 1939, the discovery of insecticidal properties of DDT (Dichlorodiphenyl trichloroethane) by Paul H. Mueller changed the scenario of pest management. During World War 2 DDT was extensively used to prevent epidemics of several insect vectored diseases such as yellow fever, typhus elephantiasis and malaria. This drew attention to the possibilities of more synthetic insecticides and as a result the use of pesticides in various arenas soared from 1940-60, complete reliance on pesticides intensive pest management was leading agriculture on a "pesticide treadmill". The overreliance on synthetic pesticides from late 1940s to mid-1960 was referred to as "Dark Ages" of pest control. The cheapness and effectiveness of synthetic insecticides threw natural compounds into shade. But very soon other shades also began to appear. In 1962, the appearance of book "Silent Spring" (by Rachel Carison) showed that pesticide residues were building up in ecosystem with detrimental effects on wildlife and beneficial insects. Due to extensive and intensive use, misuse and abuse of insecticides the following problems were becoming prominent and intolerable (ecological backlashes):

1. Development of insecticide resistance- many insects started developing resistance against pesticides which increased the cost of management.
2. Due to killing of natural controlling agents, the phenomenon of pest resurgence became more evident.
3. Also minor pest were achieving the status of major pest i.e. secondary pest outbreak due to significant decline in its natural enemy (predators and parasites).

4. Ecological imbalance due to poisoning of all the realms of environment.
5. Increase in the concentration of hydrochlorinated insecticide in food chain.
6. Intolerable residues on the food made the food obtained after such treatment uneconomical as it became unfit for consumption and unfit for exports due to high toxic residues.
7. Killing and harmful effects on Non target organisms became more prominent (like birds, fishes and other wildlife).

Overviewing these effects, there was an utmost need for the development of environmentally sound management practices. This lead to the idea of Integrated Pest Management (IPM) . A panel of experts put the concept of IPM in 1968. IPM as defined by FAO is a system which in consideration with the present environment and pest population dynamics, integrates all the sustainable techniques of pest management as compatible a manner as possible and maintain the population of pest below the level which can cause economic damage (i.e. below economic injury level). The approach is to minimize the dependence on insecticides and maximize the use of ecofriendly methods so as to cause minimum damage to the environment. Botanical pesticides, thus is an very important component of IPM as

1. They are easily degradable.
2. Don't affect non target organisms, natural controlling agents such as predator, parasites.
3. Don't form residues
4. And has no harmful effect on humans as they are very specific in action.

Botanical pesticides refer to the use of chemical or organic compound produced by plants, plant products, which have harmful effects on the growth, development and survival of insect pests. Plants are a rich source of such organic compounds.

The practice of utilizing the derivatives of plant i.e. botanical pesticides in agriculture dates at least two millennia back in ancient China, Egypt, Greece and India. Even in North America and Europe, the documented use of botanicals extend back more than 150 years before the discovery of major class of synthetic chemical insecticides (OP, carbamates and pyretheroids) in mid 1930s to 1950. It is very clear from the recent history that the chemical insecticides have essentially relegated the botanical pesticides from an important role in agriculture to a trivial position in the market among various crop protection strategies.

The total number of 20 phytochemicals is estimated to be 500000, so far only 10000 of these have been isolated. At present four major types of botanicals are being used for the control of insects. These include:

a. Pyrethrum

b. Neem (Azadirachtin).

c. Rotenons.

d. Essential oils.

Others are in limited use like Ryania, Nicotine, Sabdella. Whereas Nicotine, Rotenene, Natural Pyrethrins constitute the outstanding example of older botanicals, extracts and compounds from the Neem tree (A.indica) have emerged as the most prominent phytochemical pesticides in recent years. Among the various biologically active compounds that can be extracted from the Neem tree like- triterpenoid, phenolic compounds, carotenoids, steroids, ketones; the tetranortriterpenoid azadirachtin has been the most extensively studied pesticide as 1) it is relatively abundant in Neem kernels. 2) has biological activity on a wide range of insects.

PLANT SECONDARY METABOLITES

Plants produce a large, diverse array of organic compounds that appears to have no function in growth and development. These substances are known as SECONDARY METABOLITES or secondary products or natural products. Secondary metabolites differ from Primary metabolites (amino acids, nucleotides, sugars, acyl lipids) as:

a. They have no direct roles in photosynthesis, respiration, protein synthesis etc

b. They have restricted distribution in plant kingdom.

In a seminal paper Fraenkel stressed the role of secondary metabolites as defense system against insects, pests and other natural enemies.

Though they play no role in growth and metabolism they play important ecological role in plants:

a. They protect plants against being eaten by herbivore and being infected by microbial pathogen.

b. They serve as attractants for pollinators and seed dispersing animals and as agents of plant-plant competition.

Because Of Their Ecological Role, Plant Secondary Metabolites Are Classified As Allelochemicals, A Term Coined By Whittaker. An allelochmical is defined as a non nutritional chemical produced by an individual of one species that affects growth, health, behavior, population ecology of another species. Plants produce an astonishing array of Secondary metabolites. Even a single plant species may produce an extensive pharmacopeia of recondite chemicals. Periwinkle for example contains about more than 100 monoterpenoid indole alkaloids. It has been estimated that plant kingdom synthesizes hundreds of thousands of different secondary metabolites. The no of identified compounds now exceeds 10000.

Secondary metabolites as plant defense is result of co evolution between plants and herbivores.

Plant secondary metabolites can be divided into three chemically distinct groups:

A. Terpenes

B. Phenolics

C. Nitrogen Containing Compound

Terpenes

The terpenes constitute the largest class of secondary products .the diverse substances of this class are generally insoluble in water. They are biosynthesized from acetyl coA.Terpenes are classified by no of five carbon units they contain as:

1. Monoterpenes: Contain 2 five carbon skeleton
2. Sesquiterpenes: Contain 3 five carbon skeleton
3. Diterpenes: Contain 4 five carbon skeleton
4. Triterpenes: 30 carbons
5. Tetraterpenes: 40 carbons
6. Polyterpenoids: $(C5)n$,where $n>8$

SOME TERPENES HAVE ROLE IN GROWTH AND DEVELOPMENT

Terpenes defend against herbivore in many plants. Terpenes are toxins and feeding deterrentsto many plant feeding insects, thus they appear to play important defensive role in plant kingdom and protection of agricultural crops. Examples of important Terpenes:

1. **PYRETHROIDS:** These are monoterpenoid that occurs in leaves and flowers of Chrysanthemum species show very striking insecticidal activity. Both natural and synthetic pyrethroids are popular ingredients in commercial insecticide because of their low persistence in the environment. Pyrethrum is the predominant botanical in use accounting for 80% of global botanical insecticide.

2. **ESSENTIAL OILS:** These are the mixture of monoterpene and sesquiterpene that lends a characteristics odor to the foliage .e.g Menttholin Peppermint oil and Limonenein lemon oil are monoterpenes. Essential oils have well known insect repellent properties. They are frequently found in glandular hairs and serve to advertize the toxicity of plant repelling potential. Phytophagus insects even they take a trial bite.

3. **VOLATILE TERPENES:** In corn & wild tobacco certain monoterpenes and sesquiterpenes are produced and emitted only after insect feeding has already begun. These substances prevent oviposition and kill plant feeding insects and so help in controlling further damage .These also attract natural enemies of plant feeding insects so promise a sound means of pest control.

4. **LIMNOIDS:** These are a group of nonvolatile Triterpene.Among these the most powerful deterrent to insects feeding known is Azadirachtin.It is a complex limnoid from Neem tree which is feeding deterrent to some insects at as low as 50ppm and it exerts a variety of toxic effect .It has considerable potential as a commercial insect control because of its low toxicity to mammals.

PHENOLICS

Plants produce a variety of secondary products that contain a phenol group, these are called phenolic compounds. Plants phenolics are a chemically heterogeneous group of nearly 10000 compounds .many of these serve as defense compounds against herbivores.

- The release of phenolics into soil limits the growth of other plants.
- LIGNIN a highly branched polymer of phenylpropanoid group has significant protective function in plants. Its physical toughness deters feeding by insects and chemical durability makes it relatively indigestible.
- The flavoids are one of the largest classes of plants phenolics e.g.anthocyanins, flavones etc.
- Anthocyanins are colored flavonoids that attract insects to flower and fruits by providing visual and olfactory signal.
- Flavonoids protect against damage by UV light.

- Tannins deter feeding by herbivores and it also act as feeding repellents to a great diversity of insects

NITROGEN CONTAINING COMPOUND

A large variety of plant secondary metabolites have nitrogen in their structure. This category includes well known defense against phytophagus insects as alkaloids and cyanogenic glycosides,glucosinolates.

ALKALOIDS: These are a large family of more than 15000 nitrogen containing secondary metabolites with a heterocyclic ring. Several different types including nicotine and its relative are derived from ornithine .Most alkaloids now function as defenseagainst their predators because of their toxicity and deterrence capability. Alkaloids increase in response to initial damage fortifying against further damage e.g. wild tobacco produces higher level of nicotinefollowing damage by tobacco caterpillars.

CYANOGENIC GLYCOSIDES: These are not toxic themselves but are readily broken down to give off volatile poisons; well known poisonous gas Hydrogen cyanide. When the leaf is damaged due to insects feeding on it, the cell content of different tissue mix and HCN is formed .HCN is a fast acting toxin that inhibits metalloprotiens such as iron containing cytochrome oxidase; a key enzyme of mitochondrial respiration, thus affecting physiology of insects. Thus presence of cyanogenic glycosides deters feeding by insects.

GLUCOSINOLATES: A Class of plant glycosides that break down to release volatile defensive substances, also called Mustard oil glycosides.Found principally in the Brassicaceae and related plant families, where glucosinolates give off compounds responsible for smell and taste of vegetables like cabbage, cauliflower, mustards etc. These compounds function in DEFENCE as toxin and feeding repellent. But certain insects are adapted for feeding on glucosinolate containing plants without ill effects. For example glucosinolates serve as stimulant for Cabbage butterfly for feeding and egg laying and isothiocyanates serve as volatile attractants.

PLANT PROTIENS: Certain plant protein also interfere with insect digestion, for example plants produce LECTINS, defensive proteins that bind to epithelial cell lining digestive tract and interfere with nutrient absorption .The best known anti digestive proteins in plants are protein inhibitors found in legumes, tomatoes, and other plants. After entering herbivores digestive tract they interfere with protein digestion, as a result insects suffer reduced rates of growth and development.

THE NEEM TREE, PROPERTIES AND DISTRIBUTION

The Neem tree also known by names like Indian Lilac, Margosa tree is an evergreen fastgrowing tree belonging to the order "Rutales" and family "Meliaceae". The genus Azadirachta indica was described by A. juss in 1830.

DISRIBUTION

Neem tree is indigenous to Indian Subcontinent from where is has spread to many Asian and African countries such as Pakistan, Bangladesh, Mynamar, Sri Lanka, Thailand, Indonesia, Malaysia, Singapore, Iran, Yemen, Australia, New Guinea, Nigeria, Fizi, Tanzania, Madagascar, USA, Latin America, Germany, France, Portugal, Spain and UK. It is now grown in most tropical and sub-tropical parts of the worls.

The origin of A.indica is not very clear. Some say that is has originated from Burma whereas others point it to south India. It is considered that it has originated from south-eastern and southern Asia.

In Indonesia Neem exists in low lying Northern and Eastern parts of java. In Philippines it was introduced from India, Africa. Ketkar (1967) reported about 14 million trees in India. There are more than 20 million trees available in entire India. In Africa Neem was introduced from India and is concentrated in a belt stretching across the African continent from Somalia to Mauretania. In America Neem trees are prominent in Haiti, Surinam and propagation has started in Brazil, Puerto Rico, Cuba and Nicaragua. Neem trees also grow in our neighbouring countries, Middle East, Saudi Arabia and Yemen.

HABITAT

Neem tree is a fast growing sclerophyllous tree.

- It grows well in humid to semi-humid climate.
- It thrives well at altitudes upto 700-800m above the sea level.
- Neem trees are hardy and are able to grow in severe drought condition also. They thrive well in regions with less than 500 mm annual rainfall and upto 2500 mm annual rainfall.
- Neem tree exist in poor, shallow, sandy and stony soil. It also grows in black cotton soil in India.
- Neem tree can flourish in warm to very hot climates. It grows well between 21-320c temperatures but it can tolerate upto 500c during summer.

- Ph value between 6.2-7 seems to the best for the growth of Neem tree.

CHARACTERISTIC FEATURES

- It is a fast-growing tree, reaching a height of 4-7 m during the first 3 years and 5-11m during the following 5 years.
- It begins to bear fruit within 3-5 years and becomes fully productive in the 10th year, when it may yield up to 50 kg fruit per tree per year.
- The Neem tree produces its fruits, which are the main source for its production of pesticides, on drooping panicles, usually about once a year, although two fruiting periods per year occur in certain areas (e.g. West Africa).
- A mature Neem tree produces annually 30-50 kg of fruit, but this may depend upon rainfall and soil conditions. More conservative estimates range around 20 kg per tree; 40 kg of fresh fruit yield about 24 kg of dry fruit.
- Neem has the reputation of possessing a large number of biological activities which include insecticidal, nematicidal, bactericidal, and anti-fungal. It has attracted world-wide attention due to its wide ranging capacity as a biocide.

Pesticidal constituent of neem

Neem tree is the only tree in which every part of tree produces biologically active products which has various properties such as antifeedant, deterrent, growth regulation, oviposition alteration, insecticidal properties, fungicidal properties,etc.

Though bark, heartwood, leaves, fruits of it produce these substances in various concentrations but it is the fruits specifically seeds which are of major importance. Neem seed kernels contain the highest amount of the active compound. 40-50 kg of fruit can yield about 5 kg of kernels (10% of fruit). Each seed contains about 1-3 kernels.

Till date more than 140 active principles have been identified in different parts of the tree. Insecticidal properties of Neem is due to the presence of a class of Limnoids which include compounds like Azadirachtin, Melantriol, Salanin, Mimbines, Salannol and various sulfur containing compounds. Among these Azadirachtin is the most active and predominant insecticidal compound concentrated mainly in the seed kernels. The Azadirachtin occurs in seeds at the concentration of about 0.1-0.9%. It is estimated that 20-30 kg of Neem seeds are required per hectare if 2g of Azadirachtin per kg of sed is obtained. The highest yield of Azadirachtin obtained till date was about 10g/kg of seed.

AZADIRACHTIN AND ITS STRUCTURE

Azadirachtin is a highly oxidized limnoid chemically being a tetranortriterpenoid and is the main component responsible for both anti-feedant and toxic effects in Azadirachtin. Butterworth and Morgan were the first to isolate Azadirachtin in 1968 from Neem seed. Morgan established correct molecular formula of Azadirachitn (C55H44O16). In 1971 they developed a simplified method to isolate azadirachitn by doing solvent partitioning followed by column and preparative thin layer chromatography. However its structure was determined in 1975 by Nakanishi's team through the application of new NMR methods. There were some inaccuracies in the given model. Then again renewed efforts were made by the group of Ley, Kraus, Nakanishi and they gave the correct structure by using X-ray crystallography.

A.indica produces a plethora of triterpenoids, the biosynthesis of which culminates in azadirachtin. The biosynthesis of azadirachtin starts with a steroid precursor - tetracyclic triterpene "tirucallol". Opening of C-ring followed by processing via two main levels of structural complexity i.e. furan ring formation leads to Azadirachtin.

VARIOUS PESTICIDAL FORMULATIONS

Neem insecticides which are obtained from Neem seeds contain various arelated triterpenoids in addition to the Azadirachtin. However their efficacy is related directly to the content of Azadirachtin. These compounds do possess biological activity and they add to its effects. Pure Azadirachtin was shown to be effective in the fields (Mordue et al, 1997) but the natural mixtures of azadirachtin in Neem insecticides may usefully mitigate against the development of resistance compared to azadirachtin alone (Feng and Isman, 1995).

The complex nature of azadirachtin and other sophisticated Neem constituents prevent their mass production by synthesis in the foreseeable future. The pesticidal Neem products used in practice include dried leaves, whole seed, decorticated seed, seed kernels, Neem oil, and Neem cake, remaining after extraction or extrusion of the oil from the seeds. Several Indian companies or institutions produce commercially Neem-based insecticidal formulations, such as "RD-9 Repelin" and "Wellgro", for spraying against cutworms and other insect pests in tobacco growing areas; "Nimbosol" and "Biosol" for control of whiteflies; and the products "Neemrich" and "Neemark, the latter also as an azadirachtin-enriched granular Neem formulation. In the U.S.A., the EPA hasgranted registration to "Margosan-O", an azadirachtin-enriched, concentrated Neem seed kernel extract formulation, for use on non food crops and ornamentals. Margosan-0 was developed by R.

Larson of Vikwood Botanicals Inc. at Sheboygan, WI, in collaboration with the USDA Agricultural Research Center at Beltsville, MD. The rights to this product, which contains 0.3% azadirachtin and 14%Neem oil (the 0 in the name of the product stands for oil), and has an oral toxicity in excess of 5,000 mg/kg in rats. Margosan-0 has been evaluated successfully against an extensive series of insects in the U.S.A. and Canada, Lyriornizu leafminers on ornamentals and tomatoes, cotton bugs, cockroaches and mosquitoes. Margosan-0 demonstrated highest activity against Ostriniu nubilalis , and against leafhoppers, against two species of local cotton pests, Enrias insulana and Spodoptera littoralis. Recently in the U.S.A. a further Neem formulation, developed.under the auspices of the Natural Products Institute, Salt Lake City, UT is "Azatin"(Agridyne Technologies, Salt Lake City, UT). Also, Safer Ltd., a Canadian manufacturer specializing in environmentally safe pest control formulations, developed insecticides based on Neem. Safer, however, has been acquired recently by Ringer Corp., Minneapolis, MN, which distributes Margosan-0 in the home garden market under the tradenames of "Bioneem" and "Neemesis".

Contrary to registration practices in use until now, no precise chemical descriptions of all the ingredients of Margosan-0 were required, but rather, demonstration of the biological activity and innocuousness of the whole mixture to no target organisms was used in the registration process. Hopefully such specially tailored toxicity studies will be used to judge and register Neem and similar natural products in the future. A recent report claims that the EPA has approved a Neem-based biological pesticide developed by an Indian company for use on a wide range of food crops.

MODES OF ACTION

Major modes of action of azadirachtin are:

1. Powerful IGR.
2. Feeding Deterrant.
3. Oviposition Deterrant.

These are the three modes of action of azadirachtin which make azadirachtin much sought after biopesticide in today's agriculture industry.

IGR: Azadirachtin acts as a powerful growth regulator for insects and this IGR effect is the most pronounced mode of action of Azadirachtin.

Normally IGR effect the hormonal system of insects, preventing the insects from developing into normal mature insects. This IGR property of Azadirachtin doesnot leads to immediate death of insects, pests.

Azadirachtin as an IGR

The IGR property of Azadirachtin arise due to the fact that:

Azadirachtin is structurally analogous to natural hormone Ecdysone. As Ecdysone regulates the development of insect, any disruption in its balance leads to improper development.

Also Azadirachtin interferes with the production and reception of Ecdysone at the time of insects' growth and moulting. Thus Azadirachtin in this manner block the moulting cycle resulting in the death of the insect, pest.

The main action of Azadirachtin appears to be at the release site of PTTH. The mode of action of Azadirachtin as IGR is thus an Indirect Physiological Effect. It is exerted via the endocrine system. The copora cardiaca is supposed to be the target for the Azadirachtin as is affects the PTTH, Eclosin Hormone, Bursicin Hormone release. PTTH release is inhibited rather than Ecdysine from Prothorasic gland. Thus the Azadirachtin affects the neurosecretory cells of Brain. Various experiments show that Azadirachtin doesn't directly act on Prothorasic Glands.

1. In the in vitro culture of Prothorasic (H. virescens) gland showed that the PTTH induced release of the Ecdysine was medium (Bidmon et al, 1987, Barnby and Klocke, 1990). Also it was not blocked in PTTH simulated cultured glands from M.sexta pupa penetrated with Azadirachtin in last larval instar (Pener et al, 1988). However receptivity of Prothorasic gland to PTTH was affected in H.virescens.
2. Neurosecretory proteins stained with paraldehyde in L.migratolia females when was compared with similar aged azadirachtin treated females there was an accumaulation of stainable material in corpora cardiaca of brain neurosecretory system in treated insects. Thus is appears that azadirachtin blocks release of neurosecretory material from corpora cardiac.

It can thus be concluded that Azadirachtin does block the release of peptide hormones from brain neurosecretory cell corpora cardiac complex.

Azadirachtin also exhibit IGR effect by altering the titre of Juvenile Hormone (JH). Azadirachtin affects the release of allotropins into corpora dillata hence block the synthesis and release of the Juvenile Hormone. This block leads to a rapid decrease in whole body JH titres, which is maintained for several days. Experiments prove that in M.sexta larvae, azadirachtin infection on day 0 (1.0-10 µg/ larva) results in induction of supernumerary moults (Sch et al, 1985; Beckage et al, 1988) presumably due to an inhibition and subsequent delay in JH titre. In adult female L.migratolia also azadirachtin treatment causes a rapid decrease in juvenile hormone titres with associated disturbances in oogenesis (Rembold, 1984; Rembold et al, 1987).

Thus, on a conclusionary note, the effect of azadirachtin is both dose and time dependent. It prevents both apolysis and ecdysis and thus can cause death before the moults, during the moults or delays of moult to form permanent larvae.

Feeding Deterrance: Feeding behavior is both dependent on chemical senses stimulated due to contact chemoreceptors on trasi, mouthparts and oral activity and integration of the sensory code with the CNS. Azadirachtin acts as feeding deterrant. Inhibition of the feeding behavior occurs:

There are receptors present on and around mouthparts of insects which normally respond to Phagostimulants. So azadirachtin may act by blocking the input from these receptors.

Also there are present specific "deterrent cells" in insects which prevent insect from feeding. Azadirachtin acts to stimulate these "deterrent cells" leading to feeding deterrence. Many experiments were done in this regard.

Using different concentration of sucrose and azadirachtin, either singly or together, the neurophysiological responses from medial and lateral sensillia styloconica of maxillae showed different group of receptors are receptive to sucrose (sugar cells) or azadirachtin (deterrent cell) in S.exempta and M.brassicae in most of the cases, the rate of firing of sugar sensitive cells were reduced in presence of both chemicals (Simmonds and Blaney, 1984). Such an interaction was also found in P.brassicae. This leads to a reduced or complete inhibition of feeding.

Direct mode of action: Incorporation of azadirachtin results in direct toxic effect after ingestion. Azadirachtin prevents the secretion of Proteolytic enzymes and thus significantly impair ability of insects to digest and absorb nitrogenous food. When azadirachtin is ingested it can result in the disfunctioning of gut, as a result of which midgut epithelial cells become round. Swelling of cells and organells occur with some vacuolization and cell burst resulting in necrosis (as observed in S.gregarea and L.migratolia Naseruddin and Mordue (Luntz), 1993a; Cottee, 1984). There is also reduction in the regenerative cells and increase in the connective tissue layer with some invading heomocytes. This would lead to disruption of enzyme secretion and nutrient absorption.

Also the antifeedant effect can be attributed to the action of azadirachtin on the peristaltic movement of gut wall. The gut of treated insects lack tone, midgut to hindgut junction becomes flaccid and co-ordinated peristalsis is lacking which leads to antifeeding behavior.

EFFECT OF AZADIRACHTIN ON INSECTS

Azadirachtin is a classical example of a natural plant defence chemical affecting feeding. Antifeedancy is the major insecticidal effect of Azadirachtin. Antifeedant effect in insect pest on application of Azadirachtin is divided into two main categories:

1. **Primary Antifeedancy:** It refers to the deterrence of feeding in insects. Primary Antifeedancy is also called Gustatory antifeedancy. It can be defined as the inability to ingest resulting from the perception of antifeedant at a sensory level (Schmutterer 1985). Insects fail to eat treated crops and as starvation ensued results in the death of insects.

2. **Secondary Antifeedant effect:** It refers to the non-feeding after the ingestion of treated plant. Secondary antifeedancy is also called Non-Gustatory antifeedancy. It can be defined as the reduction in food consumption and digestive efficiency subsequent to and as a consequence of ingestion, application or injection of antifeedant (Schmutterer, 1985).

Experiments conducted in the past in this regard by various persons:

The first detailed experiment was conducted in S. gregaria (desert locusts) in India. Insects from different orders show marked difference in their response to azadirachtin.

1. Lepidopteras showed extreme sensitivity to azadirachtin and depending upon species, effective anti-feedance was observed from less than 1 to 50 ppm.
2. Hemiptera (Homoptera), Coleoptera are less sensitive to azadirachtin with 100 % antifeedancy observed at 100-600 ppm.
3. However, in Orthoptera wide range of sensitivity has been observed.

Reed and Pierce in 1981 tested the repellant effect of Neem extract to striped cucumber beetle (A.vittateim), by cutting leaves and dipping them in extract solution and placing them in a dish with untreated leaf pieces. When 5 fasting beetles were placed in a dish, 0.1 % azadirachtin gave protection for atleast three days.

The intake of food by various homopteran insects Nilaparvata lugens, Nephotettix virescens was significantly reduced on rice plants sprayed with 1-50% emulsion of Neem oil. In green rice leafhopper, N.virescens feeding on the phloem of neem oil treated plants (1.25-10%) was significantly less than of solvent treated control plants, whereas xylem feeding increased. Hemipteran insects feeding on tobacco seedlings which had been systemically treated with 500 ppm azadirachtin, were shown initially to feed normally but, after termination of the initial feed, the interval prior to the next subsequent feed was significantly increased and feeding activity thereafter was suppressed (Nisbetet al. 1993). When azadirachtun was impregnated on discs at a concentration of 0.1-10 ppm, S. littoralis(African cotton leafworm),Spodoptera frugiperda(J.E. Smith) (fall armyworm),Heliothis virescens(F.) (Tobacco budworm) andHelicoverpa armigera(Hüb.) (Old world bollworm) showed significant behavior response and are prevented from feeding on the discs dependent on species (Blaney et al. 1990, Simmonds et al. 1990, Mordue (Luntz) et al. 1998)

Insects from different Orders differ markedly in their behavior responses to azadirachtin. Lepidoptera are extremely sensitive to azadirachtin and show effective antifeedancies from <1-50 ppm, depending upon species. Coleoptera, Hemiptera and Homoptera are less sensitive to azadirachtin behaviorally with up to 100% antifeedancy being achieved at 100-600 ppm although there are some aphid species which also show behavioral sensitivity e.g. strawberry aphid. The Orthoptera show an enormous range in sensitivity from S. gregaria (a polyphagous species which has chemoreceptor finely tuned to many plant secondary compounds) to Locusta migratoria(L.) (a graminaceous species which does not have chemoreceptors tunes to feeding deterrents) to the extreme insensitivity of Melanoplus sanguinipes(Fab.), the North American Plains grasshopper which is an evolutionary sense has never encountered A. indica and has no chemoreceptors responding to azadirachtin. Such 'primary' (or gustatory) antifeedancy - 'the inability to ingest resulting from the perception of the antifeedant at a sensory level' (Schmutterer 1985), is responsible for crop protection in several species of Lepidoptera and S. gregaria.Desert locusts (S. gregaria) are very sensitive to azadirachtin and fail to feed on sugar impregnated discs when the compound is present at concentrations of 0.01 ppm and above (Mordue (Luntz)et al. 1996). Azadirachtin sprayed onto barley seedlings infested withS.gregarianymphs protect plants at low doses (2 ppm) (Nasiruddin & Mordue (Luntz) 1993). S. littoralis(African cotton leafworm), Spodoptera frugiperda (J.E. Smith) (fall armyworm), Heliothis virescens (F.) (tobacco budworm) and Helicoverpa armigera (Hüb.) (old world bollworm) also respond behaviorally to low concentrations of azadirachtin and are prevented from feeding on discs impregnated with the compound at concentrations of 0.1 - 10 ppm dependent upon species (Blaneyet al. 1990, Simmondset al. 1990, Mordue (Luntz)et al.1998).

The antifeedant effects observed in these species are highly correlated with the sensory response of chemoreceptors on the insect mouthparts (Mordue (Luntz) et al.1998). Feeding behavior depend upon both neural input from the insects' chemical senses (taste receptor on tarsi, mouthparts and oral cavity) and central nervous integration of this 'sensory code'. Azadirachtin stimulates specific 'deterrent' cells in chemoreceptors and also blocks the firing of 'sugar' receptor cells, which normally stimulate feeding (Blaneyet al.1990, Simmondset al. 1990, Mordue (Luntz)et al.1999). This results in the starvation and death of these species by feeding deterrency alone.

In most other species of phytophagous insect however crop protection results from a combination of antifeedancy and physiological effects resulting from ingestion of azadirachtin. These physiological effects include 'secundary'antifeedancy whereby feeding is reduced post-ingestively. These "secondary" antifeedant effects include 'a reduction in food consumption and digestive efficiency subsequent to, and as a consequence of, ingestion, application or injection of the antifeedant' (Schmutterer 1985).

Such secondary antifeedant effects result from the disturbance of hormonal and/or other physiological system e.g. movement of food through the gut, inhibitions of digestive enzyme production, effects on the stomatogastric nervous system etc. Mordue (Luntz)et al.1985, Koul & Isman 1991, Timmins & Reynolds 1992, Trumm & Dorn 2000). For example locusts injected with azadirachtin, which by-passes the taste receptors, show a reduced ingestion of food as seen by faecal pellet production (Nasiruddin & Mordue (luntz) 1993). Also aphids which had fed on artificial diets containing much lower concentration of azadirachtin (25 ppm) exhibited no signs of primary antifeedant effects during an initial 24h period of access to the diets but their feeding rate fell dramatically in the subsequent 24h period (Nisbetet al.1994).

A consequence of interrupted feeding activity can be an effect on the ability of insects to transmit pathogens. Aphids require extended feeding periods to acquire persistently-transmitted luteoviruses (e.g. potato leafroll virus, PLRV) from plants. Treatment of PLRV-infected tobacco plants with azadirachtin reduced sustained feeding byMyzus persicae(Sulzer) (peach-potato aphid) and reduced the ability of aphids to acquire and transmit PLRV. However, azadirachtin does not always reduce the spread of plant virus diseases by aphids. Treatment of uninfected seedlings with the same concentrations of azadirachtin (500 ppm) failed to prevent them from becoming infected when viruliferous aphids fed on them (Nisbetet al. 1996a). The successful infection of a plant with luteoviruses is dependent upon the transfer of aphid saliva to the plant, a process which may be brief by comparison with the time required for virus acquisition by the aphid, and is not overcome by the presence of the antifeedant. Similarly, azadirachtin failed to protect seedlings from infections with a non-persistently transmitted potyvirus (potato virus y) from viruliferous aphids (Nisbet 1992).

Effects on Physiology

The physiological effects of azadirachtin are much more consistent than the antifeedant effects, and result from interference with growth and moulting, interference with reproduction and interference with cellular processes. In all species tested dose response effects can be seen as reduced growth, increased mortalities, abnormal moults and delayed moults. These effects are related to disruption of endocrine system controlling growth and moulting. The moulting effects are due to a disruption in the synthesis and release of ecdysteroids (moulting hormone) and other classes of hormones and this can be demonstrated by accurately timed injections of azadirachtin into the haemoulymph of vthinstar nymphs ofL. migratoria(Mordue (Luntz)et al.1986). Measurements of haemolymph ecdysteroid levels by radioimmunoassay (RAY) revealed the normal peak of hormone release at day 8 of an 11-day instar. In those insects injected with azadirachtin before hormone release, ecdysteroid release is blocked entirely and the insects die before the moult after an extended instar; in those injected at the start of ecdysone release, the peak is delayed and its decline slow down. This prevents the release of eclosion hormone which controls the motor programme of eclosion or moulting and these insects also die before the moult. Finally, if injected at the peak of ecdysone release moult initiation proceeds but the insects die during ecdyses unable to swallow enough air to extricate themselves from the old cuticle.

The physiological effects of azadirachtin can be categorized in two ways:

i. Indirect Effects- Exerted via the endocrine system. The neurosecretory system of the brain affected by azadirachtin, which causes a blockage of the release of morphogenetic peptide hormones e.g. PTTH (prothoracicotropic hormone) and allatostatins. These control the function of the prothoracic glands and the corpora allata respectively. Moulting hormone (â-hydroxyecdysone) from the prothoracic glands in turn controls new cuticle formation and ecdyses (the act of extrication from the old cuticle) whereas juvenile hormone (JH) from the corpora allata controls the formation of juvenile stages at each moult. In the adult both hormones can be involved in the control of yolk deposition in the eggs. Any disruption in these cascade events by azadirachtin results in the many various but well-defined effects seen as moult disruption, moulting defects and sterility effects.

ii. Direct Effects- On cells and tissues. Azadirachtin is taken up into cells and causes inhibition of both cells pision and protein synthesis. Such effects are seen in flaccid paralysis of muscles, midgut cells necrosis and loss of nidi (regenerative cells) of the gut and lack of midgut enzyme production.

The sum total of the physiological effects of azadirachtin is consistent throughout species when compared to antifeedant effects. An ED50of around 1 mg/g body weight is seen though the many insects species tested (Mordue (Luntz) & Blackwell 1993).

Effects on Reproduction

When the primary antifeedant properties do not operate due to low sensitivity of chemoreceptors or are circumvented by injection or applying the compound topically, azadirachtin can be shown to cause profound effects on the reproductive process of both male and female insects. For example, inL. migratoriaazadirachtin inhibits both oogenesis and ovarian ecdystereroid synthesis so preventing oviposition (Rembold & Sieber 1981). Aphids are insensitive to the primary antifeedant effects of azadirachtin below 100 ppm, although secondary antifeedant effects are observed (Nisbetet al.1994). When female aphids are fed on diets containing low concentrations of azadirachtin (5ppm), their fecundity decreases dramatically within 48h of feeding and, if they were fed in diets containing more than 10 ppm azadirachtin any nymphs which were produced were non-viable (Mordue (Luntz)et al.1996).

Male reproduction is also affected by azadirachtin. Injection of maleO. fasciatuswith 0.125 mg per insect severely reduces male potency as seen by an 80% reduction in the fecundity of normal females when mated with treated males (Dorn 1986). Testes dimensions of male desert locusts injected with low concentrations of azadirachtin during their development were significantly reduced and the meiotic processes which are responsible for the production of mature sperm in adult males were interrupted. Blockage of cell pisions was shown to occur prior to metaphase (Lintonet al.1997). Metaphase is the stage of cell pision at which microtubules form the spindle apparatus prior to the physical separation of homologous pairs of chromosomes to opposite at this stage in cell pision suggest that cell microtubular events may have been affected by azadirachtin (Mordue (Luntz), Mordue & Nibet-unpublished).

UNDERSTANDING THE EFFECTS OF AZADIRACHTIN ON INSECTS

The primary antifeedant effects Lof azadirachtin on insects are produced by the stimulation of specific deterrent chemoreceptors on the mouthparts together with an interference of the perception of phagostimulants by other chemoreceptors (Mordue (Luntz)et al.1998). The secondary effects on feeding, developmental and reproductive disruption are caused by effects of the molecule directly on somatic and reproductive tissues and indirectly through the disruption of endocrine processes. Research is now being carried out to understand the effects of azadirachtin at the cellular level in insect tissues.

In mature adult maleS. gregaria, a tritium-labeled azadirachtin derivative, was shown to bind specifically to several tissues but the most intense binding per unit of protein was in preparation from testes. This binding was almost (kd8.7nM) and essentially irreversible (Nisbetet al.1995). Localization of the binding by autoradiography revealed preferential binding in the testes follicles, localized on the tails of developing sperm. This binding was therefore associated with one of the sub-cellular components of the developing sperm tail; membrane, axoneme or mitochondrial body (Nisbetet al.1996b).

Sub-cellular fractionation of Sf9 cells (captured insect cells derived fromS. frugiperda) incubated with dihydroazadirachtin during logarithmic growth phase revealed high affinity specific binding to the nuclear fraction of the cells (Nisbetet al.1997). A comparison of binding of tritiated dihydroazadirachtin to these two insect tissues shows specific, time-dependent, saturable high affinity binding in both tissues, with many similarities in binding characteristics. Preliminary characterization of the binding sites has indicated that it is proteinaceous, heat-labile and may be associated with cellular RNA (Mordue (luntz)et al.1999). Unsuccessful attempts to solubilise the protein and extract it for identification by ligand binding assays suggests that its 3-dimensional integrity within membranes is essential for its activity.

Azadirachtin prevents the proliferation of Sf9 cellsin vitroand alters both the protein content and abundance in those cells (Barry, Sternberg & Mordue (Luntz) unpublished), Rembold & Annadurai 1993). It therefore appears from these observations that azadirachtin operates at the cellular level by disrupting protein synthesis and secretion events and, more fundamentally, at the molecular level by altering or preventing the transcription of proteins expressed during and/ or translation of proteins expressed during periods of rapid protein synthesis e.g. in piding cells or cells forming new assemblages of organelles or cytoskeleton. Ongoing studies to fully characterize the azadirachtin binding sites are presently being carried out using insect cell lines.

PREVENTION OF AFLATOXIN PRODUCTION

Afurther unexpected, non-entomological finding is based on the work of Bhatnagar and coworkers at the USDNARS Southern Regional Research Center,

New Orleans. Since Neem has also antimicrobial and antifungal properties, Bhatnagar investigated the effect of Neem leaf extract on the growth of two aflatoxigenic (aflatoxin-producing) strains of the fungi Aspergillus fravus and A.parasiticus, as well as on aflatoxin production. Nonvolatile, somewhat heat-labile Neem leaf constituents did not affect fungal growth (i-e., mycelial weight), but completely inhibited aflatoxin production. This was demonstrated not only in a fungal growth medium in submerged culture, but also under in vivo conditions in developing cotton bolls. Injection of Neem leaf extract into artificial 3 mm deep surface holes on locules of developing bolls (30 days postanthesis) followed by an A. fravus suspension 48 h later, again did not affect fungal growth; however, the seeds from the locules exhibited a higher than 98% inhibition in aflatoxin production. It was demonstrated by in vitro studies that aflatoxin biosynthesis was irreversibly inhibited in the mycelia of both fungi by Neem constituents; removal of mycelia from exposure to the leaf extracts did not restore aflatoxin synthesis. Margosan-0 was equally efficient in inhibiting aflatoxin synthesis in developing cotton bolls.

PREVENTION OF VIRUS TRANSMISSION

Neem seed formulations effectively prevent or reduce transmission of rice tungro, ragged stunt and other virus diseases on rice by the green leafhopper, Nephoffefixv irescens, or the brown planthopper, Nilaparvata Zugetz

Differential Effects in Insects and Non-Target Organisms

In order to fully understand the mechanisms by which azadirachtin operates, the differential effects of azadirachtin must be distinguished:

i. In insects, to help decide which are the significant lesions involved in its mode of action.

ii. In other non-target organisms e.g. vertebrates to make quite certain that the margin for insecticide use is real and defined.

Two examples here related to firstly the effects of azadirachtin on locust excretory mechanisms and secondly to the effects on vertebrate cultured neurons. Studies with tritiated dihydroazadirachtin had indicated that azadirachtin accumulated in high amounts in Malpighian tubules, the excretory organs of insects (Remboldet al. 1988). Such concentrations must be associated with excretion of azadirachtin but also may be associated with its mode of action, it has been shown that azadirachtin reduces both basal and diuretic peptide-stimulated urinary secretions in locusts, and that the reduction in stimulated urine levels is induced through inhibition of cyclic AMP (cAMP) - regulated processes (Mordue (Luntz), Coast, Mordue & Nisbet unpublished). This reduction however, occurs in the presence of azadirachtin at mM levels only, with the threshold response being close to this, i.e. at levels some 1000 fold less sensitive than more established azadirachtin effects (e.g. Rembold & Annadurai 1993). Azadirachtin treated insects do become slightly bloated whit time post-treatment (Cottee & Modue (Luntz) 1982, Nasiruddin & Mordue (Luntz) 1993) presumably as a result of lesions to the Malpighian tubules, however it is very clear that lack of diuresis by the cAMP secondary messenger cascade is not the main mode of action of azadirachtin.

MAMMALIAN TOXICITY AND EFFECT ON BENEFICIAL ORGANISMS

So far, no indications of mammalian toxicity of Neem, per 0sor by dermal treatment, have been found in the USAand Germany with carefully prepared Neem seed kernel extracts under conditions of rigid test procedures; such preparations of Neem, unless misused, can be considered safe for humans. It should be mentioned, however, that infants in Malaysia and India which had been fed per 0s large doses (5 ml or more) of Neem oil as a home remedy against minor ailments, developed severe symptoms of poisoning within hours of ingestion. These consisted of vomiting, drowsiness, metabolic acidosis and encephalopathy. A4-month-old Indian child fed twice 12 ml Neem oil (on two successive days)"for cough", died 12 days later. It was assumed that the oil may have been involved in the etiology of "Reye'ssyndrome" , which, in turn, is probably due to a synergistic effect between aflatoxins contaminating the oil samples used and meliatoxins present in the oil]. This rather specific problem of Neem oil fed to infants in southern and southeastern Asia was reviewed by Jacobson. Other parts of the Neem tree may, however, be toxic to warm-blooded animals, e.g., Neem leaves to sheep], goats and guinea pigs. Neem has been shown to be outstandingly safe to beneficial organisms. Honeybees, parasitizing insects such as parasitic wasps, and predators such as spiders, earwigs, ants and predaceous mites, are not harmed by azadirachtin and Neem products .

This is due to the lack of contact toxicity in most cases, the lack of a direct ovicidal effect, and the absence of toxicity against nonphytophagous adult insects. Neem products are, therefore, very selective, although they have a rather broad spectrum of activity. Azadirachtin-containing insecticides act first as oral (stomach) poisons. In some cases, for instance in the soft-skinned larvae of Leptinotarsa decemlineata,the insects also react after dermal contact with the active principle (Steets, 1975). The death of the target insects is dose-dependent. It usually occurs a few days after application of Neem pesticides, but in extreme cases the larvae may live up to several weeks when they become unable to moult.' In such "permanent larvae" the imaginal discs and parts of the epidermis are destroyed, for instance in Epilachna varivestis(Schliiter and Schulz, 1984). Azadirachtin, alcoholic and aqueous extracts of Neem seeds and enriched formulations have, according to all tests carried out up to now, no oral or dermal toxicity to mammals. Neem flowers and leaves are eaten as a vegetable in Asia (India, Burma, Thailand).Honeybees under practical conditions are also not endangered (Schmutterer and Holst, 1987). In addition, important natural enemies of pests, such as spiders, earwigs, ants and some parasitic wasps are only slightly or not at all harmed (Hellpap, 1985; Mansour et al.,1987), in some cases even favoured (Saxena et al., 1981; Joshi et al., 1982). This is because of the lack of contact toxicity in most insects, the lack of ovicidal effect and the lack of or low effect against non-phytophagous adult insects. Hence, Neem products are quite selective.

DEVELOPMENT OF RESISTANCE

To date, there is no reported case of resistance to Neem, and standard procedures of selection for resistance, e.g., in Plutella xylostella, the diamondback moth, for 42generations have not led to resistance. This was true even in a deltamethrin-resistant strain. It must be stressed that P. xylostella is a problem insect, which rapidly developed high-level resistance in the field to modern synthetic pesticides, including both photostable pyrethroids and acylureas. Non evolution of resistance to Neem formulations may be due to their being mixtures of various, often related compounds. The latter obviously have several and different modes of action, accounting for manifold activities, such as antifeeding and repellent effects, including oviposition repellency; growth and development regulation; sterilizing effects, inhibition of oviposition and/or reduction of fecundity proper, and, in certain species, disruption of mating. In addition, Neern products sometimes exert interesting physical effects on insects (e.g.,autotomy induced by Neem oil in adult locusts) and occasionally even conventional-type toxicity.

Neem and Azadirachtin in Insect Control

The complexity of the molecular structure of azadirachtin precluded its synthesis for pesticide use. Extracts of Neem seeds containing azadirachtin together with several structurally related molecules have formed the basis of Neem usage in insect control (Isman 1997). Future approaches may also include the production of azadirachtin for insect control byin vitrotissue cultures of Neem (Allanet al. 1994, 1999). Neem insecticides are effective mainly as insect growth regulates and sterilants, against a broad spectrum of pest insects. Crude Neem extracts have been used at a local, small-farm level for some time in countries where Neem grows indigenously or where plantations have been established. In the major western countries of the world such as the USA and Canada and in Europe few commercial

Neem insecticides have reached the market place to date. Progress has been hampered by lack of supplies of Neem kernels of known azadirachtin content, by lack of standardization of formulated products, by cost of the product and by lack of regulatory approval of the complex mixture of compounds found in neem extracts. Until recently these problems had meant that Neem insecticides had not generated much impact on the marketplace. Times, however, may well be changing.

With the resolution of many of the problems of supply and standardization, the full regulatory approval of Neem insecticides by the USA and now in Germany for use on potatoes, apples and tomatoes, much field data is being generated which are establishing Neem insecticides as viable alternatives to more conventional approaches, particularly in integrated pest management system. Now that it is realized that disruption of growth and reproduction rather then antifeedancy are the main characteristic of pest control, Neem is being used in the field at lower concentrations than those originally recommended (>100 ppm ai). Treatment of artificial diet with levels as low as 5 ppm or 0.25 ppm azadirachtin have been shown to significantly reduce reproduction output in M. persicae(Mordue (Luntz)et al. 1996), and feeding growth and development inS. littoralis(Martinez & van Emden 1999) respectively. The value of low concentrations of Neem in pest control has generated research into combined approaches using both neem and beneficial species. In the laboratory using M. persicaeand its parasitoid Encarsia formosa 5 ppm azadirachtin treatments of leaf discs together with E. Formosa produce additive effects compared with either approach separately and can entirely prevent nymph production of M. persicae (Fig. 5) (Sugden, Armb & Mordue(Luntz) unpublished). In the field and in more complex laboratory situations, however, such results are more difficult to demonstrate. It would appear that there is a fine line between the level of azadirachtin required to affect the pest and the level which will not affect the parasitoid or predator (Belmainet al. 2000, Pereraet al. 2000, Raguraman & Singh 2000, Simmondset al. 2000). Such integrated approaches to pest control however are an encouraging way forward for the use of Neem pesticides.

Neem pesticides may also have a useful role to play in resistance management. It has been demonstrated that the effects of Neem in reducing levels of detoxification enzymes (due to its blockage of protein synthesis) may make insecticides more effective in resistant strains of insect (Lowery & Smirle 2000). Also, it has been shown in Bt resistant strains of Leptinotarsa decemlineata Say, the Colorado potato beetle, that 0.25% Neemix combined with Bacillus thuringiensiscan act as a resistance breaking compound (Trisyono & Whalon 2000). In this instance depending upon the resistance mechanism, the Neem effects may be due also to blockage of enzyme production, or to the reduced midgut cell turnover rate (Nasiruddin & Mordue(Luntz) 1993).

Summarizing the results with neem products under field conditions, simple aqueous and alcoholic, as well as enriched, formulated products have a high potential for pest control especially in developing countries where the raw material is present in abundance.However, to succeed, certain strategies must be followed, because the application of neem products differs from that of most synthetic compounds.

They are as follows:

1. As Neem products are ultra-violet sensitive stomach insecticides, the target insects must take them up as soon as possible during feeding; the more active material they consume the better. The application of Neem products should therefore coincide with the most active feeding phases of the target insects.
2. Neem products must be applied against the most sensitive larval/nymphal instars of the target insects, as there are also remarkable differences in sensitivity during metamorphosis.
3. Because of their delayed effect Neem products may be unsuitable if no further damage to treated plants is tolerable and if no insects should be present on plants during marketing. This may sometimes be the case in industrialized countries. Summarizing the many-sided results presented in this paper, it can be said that perhaps unique mode of action of azadirachtin, which means its controlling effect on insects hormones, especially ecdysone, and the favorable toxicological and selective properties of Neem products provide a basis for a new promising way of environmentally sound pest control with biorational pesticides within the framework of integrated pest management. Due to the longer residual and systemic effects pesticides based on Neem are more suitable than most juvenoids

Azadirachtin from Neem effects insects in a variety of different ways: as an antifeedent, insect growth regulator and sterilant. As antifeedant sensitivity varies greatly between insects the overriding efficacy of Neem insecticide use lies in its physiological toxic effects. An understanding of the physiological effects of azadirachtin in Neem has been reached and biochemical approaches have begun to define its mode of action at the cellular level. Further work is however required to fully understand its mode of action. It is now accepted that Neem insecticides have a wide margin of safety for both user and consumer. Increasing knowledge of how to use neem insecticides in the field is proving a solid base from which successful market penetration should be achieved.

PESTS AND DISEASES

Neem trees are generally pest-free, due perhaps to the presence of azadirachtin and other insecticidal compounds. However, Neem plantations have been badly damaged by a scale insect (Aonidiella orientalis) in Africa, and to a lesser extent in India (NRC 1992). Certain species of ants, moths and bugs are also known pests of Neem (NRC 1992). Live specimens are susceptible to borers and termites (Hearne 1975).

Pest Analysis

PEST analysis of any industry sector investigates the important factors that are affecting the industry and influencing the companies operating in that sector. PEST is an acronym for political, economic, social and technological analysis. Political factors include government policies relating to the industry, tax policies, laws and regulations, trade restrictions and tariffs etc. The economic factors relate to changes in the wider economy such as economic growth, interest rates, exchange rates and inflation rate, etc. Social factors often look at the cultural aspects and include health consciousness, population growth rate, age distribution, changes in tastes and buying patterns, etc. The technological factors relate to the application of new inventions and ideas such as R&D activity, automation, technology incentives and the rate of technological change.

It is very important for any company to look for pest analysis before entering into any foreign market. The PEST Analysis is a perfect tool for managers and policy makers; helping them in analyzing the forces that are driving their industry and how these factors will influence their businesses and the whole industry in general.

Political environment can affect business operation either positively or negatively depends on government policy and regulation in one country. Political environment covers about political stability, form of government, level of freedoms, incentives to foreign investors, competition law and policy and employment law (Morrison, 2011) while Wall et al (2010) pointed macro-political risk that affect all firms in the country and micro-political risk that affect only specific firms, industries or type of ventures. These factors should be measured by the Walmart when entered Germany.

Germany globally is acknowledged as a country with high level of political stability. The Global Competitiveness Report 2012-2013 (world economic forum, 2012) placed Germany as the top 20 for the pillar of institutional environment. The pillar of institutional environment measures the quality of political and legal dimension. Based on those fundamental reasons, Walmart entered Germany as the first step to established business in Europe.

In reality, Walmart encountered barrier from government's policy when entered Germany which the government restrict new licenses for food and groceries retailing industries. To cope with the problem, walmart acquired two local companies that are Wertkauf and Interspar to established business in German market.

Being successful in some countries, besides in its home country of the U.S., Walmart applied its basic philosophy to offer a broad assortment with everyday lower price in Germany. The strategy, subsequently, provoked price war with local retailers, such as Aldi and Lidl. Thus, this situation led to the legal issue, which against the German laws that try to inhibit the direct competition between big retailers and small-medium enterprises.

Moreover, there is also a requirement to provide financial information for every firm, based by the German commercial code, while Walmart argued that it is not obligatory for limited partnership company. This issue also experienced by many European firms against the alteration of commercial code, and then rose to a higher level in The European Court of Justice.

Also as 24/7 store chains, Walmart feel aggrieved with working hours regulation which is only maximum of eighty hours/week and not permitted to work on Sunday and holidays.

ECONOMIC DIMENSION

The economic environment unsurprisingly is essential for international business. In particular countries, when companies plan to do business, several key economic indicators should be taking into consideration. Those indicators such as level of economic development, trends in GDP, rate of inflation, wage levels and level of unemployment, strength of currency and convertibility and rates of taxation (Morrison, 2010).

Germany as Walmart's first footprint has a solid fundamental economy. With a population of 86.5 millions (as of 2011), GDP US$ 3,577 billions, GDP per capita US$ 43,742 and accounted for 3.93% share of total world GDP, The Global

Competitiveness Report (GCR 2012-2013) considered Germany as one of an innovation driven country that is the highest stage of countries development. The report that studied and benchmarked the many factors underpinning national competitiveness among 144 countries also placed Germany as the top 6th in overall competitiveness index, and also maintained its position in rank of 30th for macroeconomic environment. Before entering Germany, Walmart's management believed that it has opportunity to grow with these strong key economic indicators.

Walmart entered Germany in 1997 when the worldwide macroeconomic downturn has just begun with high unemployment rates. Germany's economy has also contracted and continued to fall in the eight years periods of Walmart's operation before finally it announced to leave the country by the end of 2006. Although the German government tried to apply substantial tax reform to boost retail sales in the country, Walmart still suffered significant losses in Germany as also experienced by other local retailers in that flat economy period.

Acquisition of weak companies in poor location was believed to be one of the main reasons of company's losses. Wertkauf served only small part of the country in remote area, while Interspar experienced huge losses and had poor brand image. Spent expenditures on existing stores renovation, remunerated high personnel costs and experienced unprofitable sales, made Walmart unable to generate revenue until closed its operation in Germany.

Socio-cultural dimension

People's behaviour is influenced by the social culture of one country, for instance religion, cultural grouping, level of education and social class (Morrison, 2010). Walmart success story in Canada and Mexico is not continued in Germany due to different culture and business environment. Lack of understanding about German culture led Walmart dealt with several issues when surviving its business.

Adjusting two or more different corporate culture create a tendency to difficulties, as Walmart attempted to integrate different companies which it had acquired and also acculturated to American culture. Wertkauf and Interspar have different style of operation and it took times and even created corporate culture shock to the existing employee to incorporate each other.

German buyers were not as welcome as customers from other countries, they felt uncomfortable if other people greet and offer assistance during shopping. This situation created culture gap where Walmart is known with its customer service based in the stores. Although customers are relatively unfriendly, they usually build a good relationship with their local retailers. Therefore, this loyalty disadvantages

Walmart as a new comer in German retail industry.

As discussed by Geert Hofstede , values in the workplace are influenced by culture. German employees are considered as individualistic workers (scored 67 out of 100, higher score means more individualistic). The employee loyalty is depends on personal preferences, and it relates to their duty and responsibility to the employer. The Walmart's German management, which bring by the company from the US, is ignorant of the importance of communication. They used English to communicate with German employees where it is not preferable for them. As a consequence, it made boundaries between employee and management.

To be success in today's competitive business environment, both companies and countries should be able to acquired technological innovation and able to support the development of technological capacity building (Morrison, 2010). Refer to the previous statement, adequate government spending on R&D, strong patent protection for the new products and inventions, energy availability and costs, transport infrastructure and costs, innovation system including skilled workforce and level of technology transfer will be crucial to strengthen the competitiveness.

Germany lies in the middle of the Europe. Proximity to other countries in the Europe was presumably the main reason for Walmart to choose Germany instead of the United Kingdom, which other american companies usually entered the Europe. This strategy enables the firm to save the cost of logistics and supply chain to expand over the Europe.

Furthermore, Walmart's IT-driven supply chain is considered as a leader in the use of efficient technology that the company invested on a private satellite network and centralised system of logistics. Thousands of suppliers incorporated into this sophisticated system but there are some retailers that failed to follow this IT system. Bringing this system into Germany had worsening the relations with suppliers since Germany's supplier were not comfortable with the Walmart centralised distribution system. Wal-Mart was also pioneering in making use of the barcode systems in 1983 (www.oxbridgewriters.com) to fasten tracking inventory and data collection. Almost three decades, the company's research and development (R&D) introduced the radio frequency identification (RFID) technology (www.cio.com) in 2005 as improvement for inventory purposes, theft and shoplifting avoidance.

Walmart has learnt the lessons of its expansion into Europe, especially Germany. The Walmart's failure is because of it lacks of comprehensive understanding on the targeted country. However, based on PEST analysis, Walmart did not do its homework to investigate in depth about the country, such as political circumstances, legal barriers, current economic situation and the outlook of the future economy, specific socio-cultural behaviour of one country, and technological advancement. The ignorance to the local regulation complicates Walmart to run its basic values. Insensitivity to economic situation also holds the firm to expand further. Thus, unpreparedness to adjust with new culture disharmonised firm internally and externally that consequently affected to sales growth and productivity. Although supply chain technology recognised as one of the firm's strength, the firm method faced refusal from some suppliers. Those lesson learned are valuable to the Walmart operation worldwide and for others such as academia and other multinational companies.

INSECTS AS FOOD

Insects represent an important food source for a wide variety of other animal species. Freshwater game fish such as trout, bass, and bream feed extensively on aquatic insects like mayflies, stoneflies, or hellgrammites. Artificial "flies" used by anglers are often made to resemble a fish's natural prey. Many toads, frogs, turtles, snakes, and lizards also consume insects as a major part of their diet. Insectivory is common among land-dwelling birds. Purple martins, barn swallows, vireos, warblers, flickers, whippoorwills, and swifts, for example, survive almost exclusively on insects. Other birds (such as egrets, quail, geese, plovers, snipes, and bluebirds) have a more varied diet, but they still derive a large percentage of their total nutrition from insects. There are even some insectivorous mammals: shrews, moles, bats, armadillos, and anteaters, for example. When other food is scarce, even foxes, racoons, skunks, and bears will turn to insects as a source of food.

Insects were undoubtedly an important source of nutrition for our early human ancestors. Even today, they are still collected and eaten by people of many cultures. In Mexico, dried grasshoppers are sold in village markets. High in protein and low in fat, they may be fried or ground into meal and mixed with flour to make tortillas. Sago grubs, the larvae of a wood-boring beetle, are considered a delicacy in Papua New Guinea. The islanders boil the larvae or roast them over an open fire. Ants, bees, termites, caterpillars, water bugs, beetle larvae, flies, crickets, katydids, cicadas, and dragonfly nymphs are among a long list of edible insects that provide nutrition for the people of Australia, Africa, South America, the Middle East, and the Far East. Indeed, Americans and other descendents of western European culture appear to be unique among peoples of the world in having such a strong cultural taboo against the use of insects as food.

Ecological ImpactAs consumers, scavengers, and decomposers, insects play a vital role in the biogeochemical cycling of nutrients. Insects help aerate the soil, improve its retention of rainwater, and enhance its tilth. They turn more soil than earthworms and redistribute nutrients within the root zone as they burrow and nest in the ground. Flies and dung beetles prevent the build-up of manure from large animals and speed up its decomposition by fungi and bacteria. Without such scavengers, the gradual accumulation of waste products from large herds of cattle and other ungulates (mammals with hooves) would soon render much of the landscape unsuitable for agricultural purposes. Faced with just such a situation, ranchers in Hawaii, Australia, and the southern United States have imported dung beetles from East Africa in an effort to accelerate the decomposition of manure on pasturelands that support large herds of livestock.As parasites and predators of other organisms, insects are part of a natural system of checks and balances that strengthens community stability and prevents explosive population growth from overrunning natural resources. So far, over 6000 insect species have been tested and released as biological control agents to fight insects and weeds that we regard as pests. But there are also countless other species that work for us as population regulators, often unnoticed until they are accidently destroyed by a natural disaster or human intervention. Indeed, human disruption of natural ecosystems is a common cause of pest outbreaks. Over half of all agricultural pests in the United States have been accidently imported from abroad: e. g. fire ants from South America, Japanese beetles from the Orient, and gypsy moths from Europe. Many of these species are not regarded as serious pests in their homelands because population growth is suppressed by native parasites, predators, and diseases.

Fig 2.2

But above all else, insects play an essential role as pollinators of flowering plants (angiosperms). These plants are dominant producers in many terrestrial ecosystems, but they cannot reproduce without insect intermediaries to carry pollen (the male gametophyte) from flower to flower. It is a close symbiotic relationship that has been shaped and polished by eons of co-evolution. The plants produce a wide array of colors, odors, rewards, and ruses to attract or seduce their insect accomplices. Each flower's architecture is specially designed to insure that insect visitors do not leave without a thorough dusting of pollen — destined, perchance, for the stigma of another flower nearby. Various flower types are adapted for luring different groups of insects — not only bees and butterflies but also many species of wasps, moths, beetles, and flies. Wild ginger, for example, produces an unobtrusive brown flower that smells like a dead rat. It is pollinated by fungus gnats, flesh flies, and other saprophytic insects.Many of our agricultural crops like almonds, apples, cherries, blueberries, cucumbers, squash, melons, etc. would be non-productive without insect pollinators. In the United States, there are more than 250,000 commercial and hobby beekeepers who manage over four million beehives and provide pollination services to crops on 20-25 million hectares of farmland each year. These services cost growers about 50 million dollars annually, but they produce a net gain of over nine billion dollars a year through increased yield and quality of agricultural products. Wild bumblebees and certain species of solitary bees (e.g., Nomia melanderi and Osmia spp.) are also important pollinators for the production of alfalfa seed and some fruit crops.

PRODUCTS AND INSECTS

Since ancient times, honey bees (Apis mellifera) have been valued for the honey and beeswax they produce. In many cultures, these products, as well as the bees themselves, are viewed with mystical or religious significance. To ancient Greeks and Romans, honey was the "nectar of the gods" — endowed with healing properties and super-natural powers. People of Asia used it as a preservative for fruits and cakes, Egyptians employed it as an embalming fluid, and doctors in Europe and colonial America applied it as an antiseptic to treat burns and lacerations. In Europe, honey was often mixed with wine or beer and could be fermented to produce a popular alcoholic beverage known as mead. For many thousands of years, honey was the only sweetener most people ever used. The crystalline form of sugar sold today was not readily available until the early 1800's when the first commercial refineries were built to extract sucrose from cane or beets. Honey is one of the beehive's principle food resources. It is produced from droplets of flower nectar gathered by worker bees. The nectar is temporarily held in the bee's foregut where enzymatic action begins to convert sucrose into dextrose (glucose) and levulose (fructose). In the hive, this nectar-enzyme mixture is transferred to waxen cells, reduced in volume by evaporation of water, and allowed to ripen into honey. The bees seal each cell with a wax cap when the process is complete. Worker bees make as many as 50,000 trips to and from the hive and visit up to 4 million blossoms in order to produce a single kilogram of honey (2.2 lbs). Large, healthy hives may average more than 25 kg (55 lbs) of honey per year. Although the market for honey is not as large or as profitable as it once was, annual U.S. production is still over 115 million kg. Most of this honey is used as a primary sweetener or as a substitute for refined sugar in baked goods. It is also an ingredient in a few cough medicines and laxatives.Specialized glands on the ventral side of a worker bee's abdomen secrete flakes of beeswax, a soft, malleable material that bees use to build the comb where honey is stored and larvae are reared. The wax has a relatively low melting point so it is easy to extract and purify with heat. Beeswax is still used commercially in the manufacture of cosmetics, candles, furniture waxes, leather dressings, waxed paper, inks, and medicinal ointments. Annual U.S. production is about 2 million kilograms.In recent years, there has been a growing market for bee pollen and royal jelly, two other products derived from honey bees. The pollen is collected when worker bees squeeze through a special screen at the hive entrance which dislodges pollen from the hind legs. Some nutritionists regard bee pollen as a "complete" dietary supplement. It is sometimes sold in health food stores, often with astonishing claims for its medicinal or restorative powers. Royal jelly is a glandular secretion that nurse bees feed to larvae of future queens. It is rich in vitamins and proteins, and is also sold for its curative properties. It has become a major ingredient in some expensive skin care products that promise to reduce wrinkles and retard aging.

Silk A silkworm, Bombyx mori, is the source of a unique natural fiber used to make silk cloth. This species now exists only in captivity where it is reared to maturity on a diet of mulberry leaves. As each larva spins its cocoon, it produces a continuous fiber of silk that is about 0.075 mm (3/1000 inch) thick and 900 to 1500 m (3000-5000 ft) in length. This "domestic" silk is highly valued for its light color and lusterous finish. Silk is also harvested from the cocoons of several other moth species (mostly Antherea spp.) that live in the forests of India, China, and Japan. These "wild" silks (Eri, Yamamai, Muga, and Tussah are the most common types) tend to be heavier, darker, and less highly valued than that of Bombyx mori. Silk is the strongest of all natural fibers. It is comparable to steel or nylon in tensile strength, but considerably more elastic. It can be dyed, spun into thread or yarn, and woven into fabrics that are warm in winter, cool in summer, resistant to wrinkling, and exceptionally light in weight. Until the discovery of nylon in 1938, silk was the only fiber strong and light enough to be used in the manufacture of parachutes, surgical sutures, and women's hosiery.Although it has largely been replaced by less expensive synthetic fibers in most manufacturing applications, there is still a strong demand for silk in today's textile industry. Commercial production is centered in the Orient (mostly China and Japan) where there is an abundant supply of mulberry leaves. Annual silk production exceeds 50,000 metric tons — most of it woven into fabric for clothing or blended with other fibers for draperies and upholstery.

Shellac Laccifer lacca, a tiny scale insect that grows on soapberry and acacia trees in India and Burma, is the source of lac, a sticky resin that forms the principle ingredient of commercial shellac. Twigs bearing the scale insects are heated to extract and purify the resin. Up to 200 insects are needed for each gram of lac (90,000 per pound). Shellac, made by dissolving the lac in alcohol, was widely used as a varnish (protective coating) for floors, furniture, draperies, photographs, playing cards, and dried flower arrangements. Alkali emulsions of shellac have been molded into electric insulators, phonograph records, and dental plates. Today, lac has been largely replaced by synthetic materials, such as polyurethane and vinyl, but it is still used as a stiffening agent in the fabrication of felt hats, leather shoes, sealing waxes, and various types of inks, dyes, and polishes. Lac is the only commercial resin of animal origin.

CochinealPainting rendered withcochineal pigment Cochineal is a scarlet pigment extracted from Dactylopius coccus, a scale insect that lives on prickly pear cacti in Mexico and Central America. First used by Aztec Indians as a medicine, a textile dye, and a body paint, cochineal was discovered by Spanish conquistadors under the command of Hernando Cortez (1519). The pigment was highly valued for the intensity and permanence of its color and became a staple of trade with Europe during the 17th century. Cochineal was very expensive because of its scarcity [150 insects are needed to produce one gram of dye (70,000 per pound)], so it was used in only the finest fabrics. During the American Revolution, British soldiers, known as "red coats", wore uniforms dyed with cochineal. Today, the textile industry has largely replaced cochineal with less expensive aniline dyes, but it is still used as a coloring agent in foods, beverages, cosmetics (especially lipsticks), and art products.

Tannic acid is a chemical compound widely used in the leather industry (for tanning and dying) and in the manufacture of some inks. Until the mid 1800's, most of the world's supply of tannic acid was obtained from the Aleppo gall, an abnormal plant growth found on oak trees in Asia Minor. The trees produce gall tissue in response to a chemical substance secreted by tiny wasps (family Cynipidae) that infest the trees. Larvae of these gall wasps live and grow inside the galls which unwittingly provide both food and shelter for the insect invaders. Today, there is no commercial market for oak galls because tannic acid can be extracted more economically from the quebracho tree, a South American member of the sumac and poison ivy family (Anacardiaceae).

3

SAMPLING, MONITORING AND FORECASTING

Pest damage potential is delineated, in large part, by an organism's ability to damage a valued commodity now and in the future. Therefore, in sampling for pests we are interested in estimating a trait of the pest population or commodity that is a good indicator of the damage potential of the pest (such as estimating pest population density or percentage of damaged plants in a cropping system). In all cases, some basic information should be considered before establishing an informal or formal sampling plan. Knowledge of the life history and identification features of the pest is critical. It is not an uncommon mistake to be looking for a wrong stage of the pest: early in the season seedling weeds or immature insects may look very different than mature weeds or adult insects. Proper identification can be crucial in determining whether or what type of management tactic should be implemented. For example, the severity of damage and control strategies differ for two-spotted spider mite and Bank's grass mite infesting corn.

Another crucial aspect of pre-sampling knowledge is an understanding of pest/plant interaction. Plant response to pest damage may not only vary by the species of pest, but also by the species or cultivar of the plant, the growth stage of the plant, and the health/vigor of the plant at the time of pest infestation. Certainly plant species vary greatly in susceptibility to a pest. For pests that do damage to a specific plant species, the pest's common name is often associated with its host plant although confusion can arise: beet curly top virus is indeed a sugar beet disease but Russian wheat aphid is a pest of wheat and barley. Plant cultivars also can vary greatly in susceptibility to pests. Examples of this plant attribute are small grain and alfalfa cultivars that are resistant to various diseases and some insects. In many cases, these resistant traits are incorporated into commonly used cultivars. Plant susceptibility will also vary, at times greatly, by the growth stage of the plant at the time of infestation. For example, pests may be very harmful when the crop is in the seedling stage but not after the crop becomes established: these are known as "seedling" pests. Crop plants often can out-compete weeds if the weeds are delayed in growth: this is a basic principal of early-season weed control. Aphids are commonly more damaging to seedling plants, but also may do some damage to older plants, particularly if the plants are stressed by lack of water or nutrients. A final aspect of pest/plant interaction to consider is the health/vigor of the plant when it is exposed to a pest infestation. Often, but not always, plants that are nutrient or water stressed are more susceptible to pest injury than if they are vigorously growing. It is worthy to note that both a lack of, and an over-abundance of, water may create an environment such that the plant is more susceptible to pest damage: many small grains tend to be more prone to pathogens during conditions of high moisture/humidity while the same plants grown in dry conditions will be prone to greater damage by Russian wheat aphid. Indeed, it sometimes appears that there is a pest for every environmental condition and plant, making it very important to have as much knowledge about pest-plant interaction as possible.

Sampling programs and protocols have the aim to assess a sampling trait in a population by inspecting a portion of the sampling "universe.148; In sampling pest populations, the sampling trait is usually the pest itself or damage of the commodity. The sampling "universe" relates to the area in which a pest management decision must be applied. Often, the sampling "universe" relates to the value of the commodity to be protected and the potential severity of the pest. We may be interested in protecting a high value crop from a severe, frequently occurring but sporadic, pest; therefore, the sampling "universe" may be an individual field in which an inspection of three or four sites per 30 acre field may be warranted. In contrast, we may be interested in protecting a low value commodity, such as rangeland, from a less severe pest; therefore, the sampling "universe" may be several square miles in which fewer inspections occur because the lessened severity of the pest does not justify a higher frequency of costly inspections. In situations of less severe, and sporadic pests on low valued commodities there is interest in developing region-wide sampling procedures that take advantage of remote-sensing automated technology in order to decrease the cost of sampling in many areas.

In pest ecology, we are usually concerned with density or frequency of occurrence of sampling traits (usually the pest but also can be plant damage). Sometimes we are only concerned with detection (some governmental inspection services that aim to detect the entry of pests into a new region). Sampling protocols may provide direct estimates of the sampling trait based on the unit of interest (plants, foot of row) or based on "catches" (traps, screens, sweeps, and shakes). Other sampling tools provide indirect estimates (such as measuring plant damage in the form of defoliation, root feeding, pierce marks, or other type of damage). These indirect measures are sometimes called indexes of pest severity. Indirect measures must be used with caution because detection of plant damage may or may not be well related with the presence of a pest population. All the sampling programs presented in this guide will involve the estimation of pest densities or frequencies of occurrence of pests (or damage in the case of indirect estimates) that relate to a plant unit, land area unit, or trap unit. Examples of estimates of pest densities include number of leaf spots per plant, number of grasshoppers per square foot, number of weeds per foot of row, number of insects caught per day in a trap, and number of larvae per sweep (using a sweep net).

ltimately, the economic value of a sampling protocol and implementation of a management tactic relates to the prevention of commodity loss because the management tactic was implemented at the appropriate time based on good estimates of pest damage potential. Of course, the savings associated with preventing commodity loss should be greater than the cost of using the sampling protocol and management tactic. Whenever available, this guide will present information on such economic thresholds: the level (usually pest density) at which the cost of sampling and implementing management tactics is equal to the cost of the expected commodity loss if a management tactic is not implemented. If the economic threshold is exceeded, it becomes economically viable to implement a management tactic. Economic thresholds are typically and most appropriately used for implementing short-term management tactics, such as chemical control and to a lesser degree regulatory and cultural controls. Chemical control is the use of chemical compounds to kill and/or reduce pest densities. Cultural control is deliberate modification of the crop environment or pest habitat to decrease pest populations. Regulatory control is the regulation of an area or commodity in order to eradicate, prevent, or manage pests and their injury to a valued commodity.

Based on the factors affecting pest-plant interaction, as well as variability in commodity price, economic thresholds should not be considered stagnant levels. Usually, economic thresholds cited in this guide will be pertinent to a particular pest stage or crop growth stage occurring under normal environmental conditions. As available, information will be presented so that the reader can modify or "build" his own threshold based on specific conditions. If a single economic threshold is given, pay special attention to the situation (pest and plant conditions) in which it is applicable. Thus, it is very important to consider that economic thresholds should be used as suggested guides that will aid you in determining whether to implement a management tactic. Keep in mind that most economic thresholds are constructed from information on commodity and control economics and an understanding of pest-plant interactions. Other factors, such as presence of beneficial organisms (wildlife, honey bees, natural enemies of pests), potential contamination of soil and water sheds, and potential human exposure to pesticides, should be considered prior to implementing a pesticide control strategy.

Other pest management strategies may be available that typically involve long-term sustainable approaches to pest control (such as biological control and cultural control). Biological control is the action of natural enemies (such as some plant-feeding species that are weed feeders, and parasitoids and predators of pests) that result in the death or suppression of pests. Cultural control is the reduction of pest populations or their effects through standard agricultural practices (such as crop rotation and use of pest resistant cultivars). If successful, these strategies usually do not require active season-to-season manipulation by a grower (such as pest control by established exotic natural enemies of an introduced pest) or are part of the normal production practices of a grower (such as growing cultivars that are resistant to a pest). Because of the nature of these control strategies, they are not implemented by standard use of economic thresholds. These sustainable approaches to pest control are admirable because they usually are very much in keeping with the goal of good pest management: use of control strategies to reduce pest populations to levels that do not cause economic injury to the commodity, without exposing humans to harmful control agents, without disrupting nontarget organism(s), and without causing other forms of harm to the environment. When available, sustainable pest management strategies will be listed in this guide. We recommend that growers consider such options for use in their operation (i.e. using a resistant cultivar) and be aware of the many beneficial organisms that contribute to a productive agriculture and healthy environment.

SAMPLING

The reason for sampling is to base pest management decisions on the best information possible. A pest management program cannot work if it is based on false premises. For example, if you mistakenly believe that the mite population in your orchard is below the treatment threshold you will not spray when you should, and this could lead to economic damage. Or, if you estimate the pest population to be above the economic threshold when it is actually below, you would spray when it was not necessary and waste money. In both cases, the fault was not in the treatment threshold or your pest management plan but in your estimate of the pest population.

done if:

feed on the fruit.

otential to increase very quickly.

onstant check on what is happening in the field. If pest
vary much, there is little use in frequent sampling.

- You need accurate estimates to determine whether natural enemies will suppress the population.

If sampling shows they will not, then you may need to take treatment.

- You need to evaluate your pest controls.

Before-and-after comparisons will show if a pesticide application or introduction of natural enemies is working and if further control is needed.

- An accurate economic injury level for the pest is available.

Although sampling will estimate pest populations, it is most useful when an economic injury level has been established to indicate when control is needed.

Accurate estimates of pest populations allow you to compare its relative density from generation to generation and decide if it needs to be controlled. Data should also be saved from year to year so you can fine-tune your management programs based on past performance.

This depends on the biology of the pest and its natural enemies. For example, the western tentiform leafminer has 3 to 4 generations per year. The potential for population increase from generation to generation is high, but a generation can take a month or more to complete. It doesn't pay to sample the leafminer more than once per generation unless you need to assess the degree of biological control. Weekly sampling of leafminer will show little change.

On the other hand, spider mite populations can complete a generation in 7 to 10 days during hot weather. Sampling should be done every 7 to 10 days so you are continually aware of what is happening before populations surpass the economic injury level.

If you take a 10-leaf sample and find 30 mites, the average number of mites per leaf is 30 divided by 10, or 3 mites per leaf. This average, or mean, gives you important information about the population, but the mites could be differently distributed among the 10 leaves. By tallying the number of leaves that have no mites, then the number with one, then the number with two, and so on, you can work out a frequency distribution. In the uniform distribution, sampling even one or two leaves gives a close estimate of the mean population. Unfortu sort of distribution is rare with insects or mites. The other extreme is the distribution. Here, most leaves have no or few mites on them, while the rem leaves have many. If you took a sample of three leaves and calculated the average, estimate could vary from 0 to 9 mites per leaf. This kind of error in estimating mea populations leads to management mistakes. Unfortunately, this sort of distribution is typical of insect populations.

To decide how many samples must be taken to accurately estimate populations, it is important to know the ratio between the mean (or average) population density and the variance. The variance is a measure of the variation in the number of insects from leaf to leaf compared with the average population density. The ratio changes as the mean population changes, but it changes in a predictable way. With a large enough collection of samples, you can plot the relationship between the mean and the variance. From this, you can predict what the variance is from an estimate of the mean population density.

Knowing the relationship between the mean and variance also allows you to predict the distribution for any given population level. Using this relationship, we can predict how many leaves need to be sampled to give a certain degree of accuracy. Even more important, it allows us to use presence-absence sampling programs. If you know what percentage of the leaves are infested at a range of densities, you can estimate the average population density from just tallying the number of infested leaves without needing to count the insects. To show how well this works, you can plot the percentage of leaves infested and the average population level calculated from counting all the mites. The problem with this type of sampling plan is that estimates of the mean population density are unreliable when more than 80% of the leaves are infested.

However, presence-absence sampling can still be used by looking for the presence or absence of some easily recognized number of insects or mites, still without counting every one. For example, using the same relationship between mean and variance, we can predict the relationship between the mean and the percentage of leaves which have, say, 2 or more mites.

Samples using higher tally thresholds (2 or more, 3 or more, etc.) are still binomial samples in that each sampling unit is put into one of two classes. Although the term "presence-absence" is still used, you are no longer truly classifying on the basis of presence or absence. Several of the sampling schemes in later sections of this manual use a tally threshold of 2 or more (e.g. twospotted and McDaniel spider mites, Typhlodromus occidentalis, western tentiform leafminer). Because estimates are unreliable when more than 80% of the leaves are infested, the highest mean that can be accurately predicted with 2 or more mites per leaf is about 10.6. That compares with 6.7 mites per leaf for the normal presence-absence sample where all infested leaves are included. If this is still not high enough, you can count only the leaves with more than 3 mites or more than 4 mites, and so on. However, beyond a certain point you save less time with the presence-absence sampling because you end up counting many of the mites anyway to make sure they exceed your tally threshold.

Sequential sampling

If you know the distribution, you can set up a sequential sample. In a sequential sample the total number of leaves (sampling units) is variable. It depends whether or not results so far show that the treatment threshold is surpassed. This can be done either by counting the number of insects or mites, or it can be combined with presence-absence sampling. For monitoring purposes, it is often easier to sample a fixed number of leaves from a fixed number of trees that will give an accurate enough estimate when the average population level is near the economic threshold.

In our contemporary society, the company is operating in a complex marketing environment, consisting of uncontrollable forces to which the company must adapt. Hence, it is necessary for the company to examine its marketing environment includes the macro and the micro marketing environment. Some people believe that the company only focuses on micro marketing environment because these are internal factors close to the company that have a direct impact on the organizations strategy. However, the others argue that the wider macro-environment changes may not be as close to the marketing firm's day-to-day operations, but they are just as important, which are Political-legal factors, Economic factors, Socio-cultural factors and Technological factors.

There are three types of environments which affects a business. They are:

(a) Internal Environment: Internal environment includes staff of the organisation, office technology, wages and finance, etc.

(b) Micro Environment: Micro environment includes the external customers, agents, distributors, suppliers, competitors of the business.

(c) Macro Environment: Macro environment basically involves the Political or legal issues, Economic factors, Socio cultural factors, and Technological factors. And these factors are known as PEST factors.

Purpose Of The Pest Analysis: There are many purposes for doing the PEST analysis like when we want to expand or start our business in the global market, then by the help of PEST analysis we can assess the perspective market, assess the competitors and standpoint for the perspective business as well.

The main purpose of analysing the PEST in macro-environment is, it is very considerable that we should identify the important factors which might affect or change the vital variables which do influence any company or organization's supply and demand levels and its costs of production factors. (Kotter and Schlesinger, 1991; Johnson and Scholes, 1993).

In the macro-environment, there are so many factors available which affects heavily the organisation activities. These factors influence the organisation's Demand and Supply levels to its cost. (Kotter and Schlesinger, 1991; Johnson and Scholes, 1993) . The Rapid Changes is now a part of the environment which gave a heavy impact on the function of the organization. A number of checklists are now in market which can be use in analysing the external and internal issues of the industry. A PEST Analysis is one of them .In the present time PEST Analysis is very essential for an organisation. What PEST Analysis does is, it categorizes the environmental factors into four parts. They are; political, economic, social and technological forces. PEST Analysis analyse the impact of each of the factor on the business. Then a result came out after the analysis. A business then use this result to take advantage of the opportunities and business also use the results to take important decisions for threats while preparing for the business and strategic plans. (Byars, 1991; Cooper, 2000).

It is very much important to understand the market growth or decline, Position and potential of the business and also the direction of operations. For knowing and understanding all this, PEST Analysis is a best tool for an organisation. The factors involve in PEST are actually a framework for analysing the actual situation. (Kotler (1998)). Also to review the strategic direction a company add PEST analysis to SWOT and Porter's five forces.

(1) STRATEGIC PLANNING: Strategic planning of a business is affected not only by internal and external factors but also by the factors available in the macro environment. Pest analysis is the tool which scans and analyse the factor of external environment which are Political, Economic, Social and Technological and called as PEST analysis. These factors affect the business as well as organisation's activities and operations. PEST Analysis helps business and organisation to know these data and to do strategic planning in a better way.

(2) MARKETTING PLANNING: The next important function which PEST helps to do for the business is Marketing Planning.

PEST analyse about the market segmentation and customer segmentation that what segment should be target by the company. It also explain what makes the company's proposition different to those of competitors, presenting details of why target customers will choose to purchase the company's product or service. All this can only be possible because of PEST Analysis.

(3) BUSINESS AND PRODUCT DEVELOPMENT: As PEST is known as an important tool for the business organization. The next purpose of the PEST is Business development as well as product development. After analysing the factors in external environment PEST shows the result to the business management .On the basis of this result management of the company decides that what to do regarding that, as in if the business require any change or development. And in the same case is the product of the company is requiring any development regarding anything. If result of the analysis displays that business or product are requiring certain changes and development then the management start taking decision and start making plan regarding that.

(4) RESEARCH REPORTS: Another purpose of the pest analysis is that it gives the research reports to the business about the external environment which may affect the business and the functions and operations of the business. The steps taken by the business after doing PEST analysis are only depends upon these Research Reports.

(5) Ensures that company is performing positively even after the change in the factors which affects business environment. (Porter, 1985).

(6) When a company decides to enter into a new market and also a new market of a new country PEST is useful in that also. How is it is, PEST helps the business to adopt the realities of the new environment and also helps to break free unconscious assumptions of the new market in a new country.

ASDA is known as "The one stop shop for every need". ASDA was started with the vision to make ASDA a one stop shop for all the needs of the consumers. ASDA is a well known supermarket chains located in United Kingdom. Consumer can find any product like food, clothes and also general merchandise items or any item according to their need. ASDA came in the market in 1999 with the well known Wal-Mart. And now ASDA is the second largest retailer in the UK, and it has been the largest subsidiary of the Wal-Mart. ASDA's scale can be appreciated by the number of stores it runs as well as how many people work for it. For example, it has 321 stores across the UK and Northern Ireland, 29 depots and more than 148,000 colleagues. These all help to generate more than £15 billion of turnover. http://en.wikipedia.org/wiki/Asda

HISTORY

- **1920:** The Asquith family owned a buttchers shop at knottingely in west York shire. The two brothers, Fred and Peter took over the business and were also the founders of ASDA. Craven Dairies ltd was formed with the help of a group of West Riding Dairy Farmers.
- **1949:** The companies got merged to establish the Associated Dairies & Farm Stores Ltd.
- **1965:** It was only in 1965 that the company got its present name when Associated Diaries came together with Asquith.
- **1998:** The Company started its online store.
- **1999:** ASDA came in the market in 1999 with the well known Wal-Mart. After it became one of the subsidiaries of Wal-Mart there were many super centres opened by both the companies jointly.
- **2003:** The company brought into the market "ASDA living" which was the first general store of the company. It sold a range of products including clothing, toys, home electronics, health products, home wares, as well as beauty products. The company has joined hands with the Compass group that has coffee shops in a few of the stores of the company.
- **2006:** ASDA introduced ASDA Essentials that stocked the products of the company.
- **2007:** The Company launched ASDA electrical that focused on electrical products. This is to compete with Tesco Direct of Tesco. (http://www.about-asda.com/faq/)

The way ASDA has branded and marketed itself, and the current economic climate, the evaluation of external factors using a PEST analysis has been vital in ASDA's success. Each external factor continues to be examined and classified in terms of whether its significance is large or small, continuous or discontinuous impact or positive or negative implication.

PEST ANALYSIS OF ASDA:

Political Factors

One of the competitive advantages of ASDA has over other competitors is its growth and marking its physical presence everywhere which also raises a concern about ASDA winning over the competition over other retailers. The risk is that quality of products and services will fall and the prices charged will go higher. Till date, ASDA has not been seen as an exploiter but should keep this in mind and should regularly make use of PEST analysis so that its dominance is not taken as exploitation.

In addition, ASDA should be aware of planning permission due to their continued expansion as planning permission is heavily regulated in UK. PEST analysis would help initially to assess the potential success of a store in new area.

Operating in a globalized environment with stores around the global ASDA's performance is highly influenced by legislative and political conditions of these countries, including the European Union (EU). For employment legislations, the government encourages retailers to provide a mix of job opportunities from lower-paid and locally-based jobs to highly-skilled, higher-paid and centrally-located jobs (Balchin, 1994).

Economic factors

Economic factors are of concern to ASDA and one of the most influential factors on the economy is high unemployment levels, which decreases the effective demand for many goods, adversely affecting the demand required to produce such goods.

These economic factors are largely outside the control of the company, but their effects on performance and the marketing mix can be deep. Although international business is still growing, and is expected to contribute greater amounts to ASDA's profits over the next few years, the company is still highly dependent on the UK market. Hence, ASDA would be badly affected by any slowdown in the UK food market and are exposed to market concentration risks.

Social/Cultural Factors

Current trends show that British customers have moved towards 'bulk' and 'one-stop' shopping, which is due to a variety of social changes. ASDA have, therefore, increased the amount of non-food items available for sale.

Demographic changes require consideration such as the aging population, increase in young population, and increase in female workers and a decline in home meal preparation. These changes naturally increase demand of few products like, increase in career minded professional will lead to more demand of ready to eat or microwaveable products. In addition, the focus is now towards operational improvements, own share of business mix and supply chain, which can drive costs out of the business. National retailers are increasingly restraining themselves to take on new suppliers (Clarke, Bennison and Guy, 1994; Datamonitor Report, 2003).

The type of goods and services demanded by consumers depends on their social conditioning, attitudes and beliefs. Nowadays, consumers are becoming more and more health conscious and their attitudes towards food are constantly changing. One example of ASDA keeping up with this change is by adapting an increased demand for organic products by adding it to its product mix.

TECHNOLOGICAL FACTORS

Development of many of the products of ASDA has been influenced by the technological advancements. The new technology benefits both customers and company resulting in an increase in customer satisfaction due to involvement of factors like personalised service, readily available goods and convenient shopping. ASDA stores utilizes following technologies:

(i) Electronic shelf- labelling

(ii) Self check out machine

(iii)Wireless devices

(iv) Intelligent scale

(v)Radio frequency identification (RFID)

PEST SAMPLING AND USING PEST MANAGEMENT TACTICS

Pest damage potential is delineated, in large part, by an organism's ability to damage a valued commodity now and in the future. Therefore, in sampling for pests we are interested in estimating a trait of the pest population or commodity that is a good indicator of the damage potential of the pest (such as estimating pest population density or percentage of damaged plants in a cropping system). In all cases, some basic information should be considered before establishing an informal or formal sampling plan. Knowledge of the life history and identification features of the pest is critical. It is not an uncommon mistake to be looking for a wrong stage of the pest: early in the season seedling weeds or immature insects may look very different than mature weeds or adult insects. Proper identification can be crucial in determining whether or what type of management tactic should be implemented. For example, the severity of damage and control strategies differ for two-spotted spider mite and Bank's grass mite infesting corn.

Another crucial aspect of pre-sampling knowledge is an understanding of pest/ plant interaction. Plant response to pest damage may not only vary by the species of pest, but also by the species or cultivar of the plant, the growth stage of the plant, and the health/vigor of the plant at the time of pest infestation. Certainly plant species vary greatly in susceptibility to a pest. For pests that do damage to a specific plant species, the pest's common name is often associated with its host plant although confusion can arise: beet curly top virus is indeed a sugar beet disease but Russian wheat aphid is a pest of wheat and barley. Plant cultivars also can vary greatly in susceptibility to pests. Examples of this plant attribute are small grain and alfalfa cultivars that are resistant to various diseases and some insects. In many cases, these resistant traits are incorporated into commonly used cultivars. Plant susceptibility will also vary, at times greatly, by the growth stage of the plant at the time of infestation. For example, pests may be very harmful when the crop is in the seedling stage but not after the crop becomes established: these are known as "seedling" pests. Crop plants often can out-compete weeds if the weeds are delayed in growth: this is a basic principal of early-season weed control. Aphids are commonly more damaging to seedling plants, but also may do some damage to older plants, particularly if the plants are stressed by lack of water or nutrients. A final aspect of pest/plant interaction to consider is the health/vigor of the plant when it is exposed to a pest infestation. Often, but not always, plants that are nutrient or water stressed are more susceptible to pest injury than if they are vigorously growing. It is worthy to note that both a lack of, and an over-abundance of, water may create an environment such that the plant is more susceptible to pest damage: many small grains tend to be more prone to pathogens during conditions of high moisture/humidity while the same plants grown in dry conditions will be prone to greater damage by Russian wheat aphid. Indeed, it sometimes appears that there is a pest for every environmental condition and plant, making it very important to have as much knowledge about pest-plant interaction as possible.

Sampling programs and protocols have the aim to assess a sampling trait in a population by inspecting a portion of the sampling "universe.148; In sampling pest populations, the sampling trait is usually the pest itself or damage of the commodity. The sampling "universe" relates to the area in which a pest management decision must be applied. Often, the sampling "universe" relates to the value of the commodity to be protected and the potential severity of the pest. We may be interested in protecting a high value crop from a severe, frequently occurring but sporadic, pest; therefore, the sampling "universe" may be an individual field in which an inspection of three or four sites per 30 acre field may be warranted. In contrast, we may be interested in protecting a low value commodity, such as rangeland, from a less severe pest; therefore, the sampling "universe" may be several square miles in which fewer inspections occur because the lessened severity of the pest does not justify a higher frequency of costly inspections. In situations of less severe, and sporadic pests on low valued commodities there is interest in developing region-wide sampling procedures that take advantage of remote-sensing automated technology in order to decrease the cost of sampling in many areas.

In pest ecology, we are usually concerned with density or frequency of occurrence of sampling traits (usually the pest but also can be plant damage). Sometimes we are only concerned with detection (some governmental inspection services that aim to detect the entry of pests into a new region). Sampling protocols may provide direct estimates of the sampling trait based on the unit of interest (plants, foot of row) or based on "catches" (traps, screens, sweeps, and shakes). Other sampling tools provide indirect estimates (such as measuring plant damage in the form of defoliation, root feeding, pierce marks, or other type of damage). These indirect measures are sometimes called indexes of pest severity. Indirect measures must be used with caution because detection of plant damage may or may not be well related with the presence of a pest population. All the sampling programs presented in this guide will involve the estimation of pest densities or frequencies of occurrence of pests (or damage in the case of indirect estimates) that relate to a plant unit, land area unit, or trap unit. Examples of estimates of pest densities include number of leaf spots per plant, number of grasshoppers per square foot, number of weeds per foot of row, number of insects caught per day in a trap, and number of larvae per sweep (using a sweep net). Examples of estimates of frequency of occurrence include the proportion of plants (or animals) that are infested/infected, or are damaged. Issues involving pest-plant interactions, economic value of the commodity, and economic cost of control must be considered to devise a good sampling protocol, whether it be a formal, well-researched, plan or a more informal plan based on good general knowledge of the following issues:

- Issues to consider in developing a sampling protocol
- What / where to sample?
- What is the destructive stage of pest (and virulence of different stages)?
- What is the susceptible stage of plant?
- What pest stage is most susceptible to management tactics?
- What are the most efficient ways to sample while maintaining good estimates of pest densities or plant damage?
- When to sample (how often)?
- What is the severity of plant response to pest attack?

- What is the epidemiology of the pest population (how quick does it increase)?
- What is the lag time necessary between sampling and implementing a management tactic?

What is the lag time necessary for a management tactic to be effective?

Ultimately, the economic value of a sampling protocol and implementation of a management tactic relates to the prevention of commodity loss because the management tactic was implemented at the appropriate time based on good estimates of pest damage potential. Of course, the savings associated with preventing commodity loss should be greater than the cost of using the sampling protocol and management tactic. Whenever available, this guide will present information on such economic thresholds: the level (usually pest density) at which the cost of sampling and implementing management tactics is equal to the cost of the expected commodity loss if a management tactic is not implemented. If the economic threshold is exceeded, it becomes economically viable to implement a management tactic. Economic thresholds are typically and most appropriately used for implementing short-term management tactics, such as chemical control and to a lesser degree regulatory and cultural controls. Chemical control is the use of chemical compounds to kill and/or reduce pest densities. Cultural control is deliberate modification of the crop environment or pest habitat to decrease pest populations. Regulatory control is the regulation of an area or commodity in order to eradicate, prevent, or manage pests and their injury to a valued commodity.

Based on the factors affecting pest-plant interaction, as well as variability in commodity price, economic thresholds should not be considered stagnant levels. Usually, economic thresholds cited in this guide will be pertinent to a particular pest stage or crop growth stage occurring under normal environmental conditions. As available, information will be presented so that the reader can modify or "build" his own threshold based on specific conditions. If a single economic threshold is given, pay special attention to the situation (pest and plant conditions) in which it is applicable. Thus, it is very important to consider that economic thresholds should be used as suggested guides that will aid you in determining whether to implement a management tactic. Keep in mind that most economic thresholds are constructed from information on commodity and control economics and an understanding of pest-plant interactions. Other factors, such as presence of beneficial organisms (wildlife, honey bees, natural enemies of pests), potential contamination of soil and water sheds, and potential human exposure to pesticides, should be considered prior to implementing a pesticide control strategy.

Other pest management strategies may be available that typically involve long-term sustainable approaches to pest control (such as biological control and cultural control). Biological control is the action of natural enemies (such as some plant-feeding species that are weed feeders, and parasitoids and predators of pests) that result in the death or suppression of pests. Cultural control is the reduction of pest populations or their effects through standard agricultural practices (such as crop rotation and use of pest resistant cultivars). If successful, these strategies usually do not require active season-to-season manipulation by a grower (such as pest control by established exotic natural enemies of an introduced pest) or are part of the normal production practices of a grower (such as growing cultivars that are resistant to a pest). Because of the nature of these control strategies, they are not implemented by standard use of economic thresholds. These sustainable approaches to pest control are admirable because they usually are very much in keeping with the goal of good pest management: use of control strategies to reduce pest populations to levels that do not cause economic injury to the commodity, without exposing humans to harmful control agents, without disrupting nontarget organism(s), and without causing other forms of harm to the environment. When available, sustainable pest management strategies will be listed in this guide. We recommend that growers consider such options for use in their operation (i.e. using a resistant cultivar) and be aware of the many beneficial organisms that contribute to a productive agriculture and healthy environment.

Biodiversity has always been in a flux with species adapting to new environments or extinction but these circumstances are all through natural occurrences and through a period of thousands of years and so allowing the dependant consumers and trophic levels to adapt. However, since the Human population began to expand hugely since the industrial revolution we have been directly responsible for the extinction of dozens of species such as the Passenger pigeon (Ectopistes migratorius) and indirectly responsible for many others. As the human population continues to grow we will require more resources and to expand our living area, in this case we have exploited the natural resources to breaking point and have expanded into synthetic products that damage the ecosystem even further. The idea that we should be working to repair the damage that has been caused by our rapid growth is a relatively new concept and has still not been accepted by many countries but efforts are being made to protect the remaining communities and ecosystems.

Humans have modified ecosystems for our benefit such as agriculture which focuses on the mass production of a selected few products for our consumption; this is extremely effective in feeding a large population while incorporating only 3 trophic levels, the producer (crops), the primary and secondary consumers (humans, livestock). This method also benefits energy, keeping any loss at a minimum especially when combined with synthetic fertilizers and pesticides. However when looking at the damage this method causes to other populations there is huge fallout, we use pesticides which kill any local insect populations, and fertilizers run off the fields and contaminate the local areas, freshwater and coastal ecosystems both biotic and abiotic. we effectively destroy large areas to accommodate these fields which remove natural habitats for many species which places a huge pressure on the ecosystem.

Using biotechnology scientists have adapted blue green algae to act as a cost-effective bio fertilizer, utilising its natural nitrogen fixation to add nutrients to crops instead of relying on the more harmful, synthetic fertilizer, reducing harmful effects of the man-made substances, using natural bio waste and recycling energy flows far more efficiently.

The human effect on the ecosystem has damaged many areas both directly and indirectly such as pollution, climate change and introduction of foreign species. In the last 50 years the global economic activity has increased 7 fold in the last 50 years and trends suggest that as our population continues to rise the ecosystem will suffer more http://www.greenfacts.org. Land and forests are in greater demand for agricultural, industrial and housing which destroy natural habitats of native species and force them out. Forest fragmentation increases the risk of extinction as the smaller a space the smaller the population it can support. Combining this with the introduction of alien species, i.e. imported from other countries and then escape into a new ecosystem, this is a major cause of the extinction to local species such as the American mink escaping into the British countryside, this non-native carnivore has posed a significant threat to fish, poultry and game birds, The introduction of new diseases that can prove devastating, Dutch elm disease has wiped out the British elm tree which then saw a huge decline in the producers for oxygen and allowed more non-native trees to grow in their place.

With recent attempts to cut our 'ecological footprint' and ease pressure on the ecosystem as much as possible we have introduced measures that manage as much of our resources as possible, Protected areas and set-aside fields have given wildlife a chance to recover its nutrients from decades of abuse while the government allows a subsidy to the farmers that do this. While protected areas such as these are a good start it is not sufficient to recover biodiversity. The restoration process is producing strong results and is evolving rapidly due to increased awareness and changes in priority, Using a holistic approach Restoration technology is now able to return certain damaged areas of ecosystems to their natural balance . (B. C. Rana

Damaged Ecosystems and Restoration) Ecological damage is now a large part of political arguments, bringing them to the public attention and changing the public attitude towards these issues as well.

"Ecosystems are already showing negative impacts under current levels of climate change â€¦ which is modest compared to future projected changesâ€¦. In addition to warming temperatures, more frequent extreme weather events and changing patterns of rainfall and drought can be expected to have significant impacts on biodiversity". (Secretariat of the Convention on Biological Diversity (2010)

Pollutants such as oil, gas fumes, and synthetic chemicals have created the most widespread impact to the ecosystem with temperatures in flux, the polar icecaps melting, acid rain and the damage to the Ozone layer. Native plants are unable to thrive in the acid rich soils making for fewer producers and less for the primary consumers to absorb this carries throughout the chain (Ann N Y Acad Sci. 2009)

Today's globalized business market, the systematic move from push to pull manufacturing, and the rise in consumer oriented economies, have led to a much more complex forecasting world (Lapide, 2006). Forecasters are being asked to create plans for expanding geographies, increased numbers of sales channels, and broader, more diverse, and shorter life cycle product lines. This complexity means that markets are more dynamic and the business environment is not stable (Lapide, 2006).

The importance of forecasting is finding in a great range of planning and decision making circumstances. It is essential to mention those perspectives that forecasting can become a useful tool for management in many departments of an organization. In marketing, a great amount of decisions can be improved significantly by connect them with dependable forecasts of market size and market characteristics (Makridakis and Wheelwright, 1989). Having this in mind for example, a company that produces and sells electrical devices should be able to forecast what the demand will be for each of its products by geographic region and type of consumer (Makridakis and Wheelwright, 1989).

In production an essential need of forecasting is the area of product demand. This relates with the both prediction of volumes mix so as the organization can plan its production schedule and organize appropriate its inventories (Makridakis and Wheelwright, 1989). Another area that the recent years have linked a lot with forecasting is finance and accounting. These departments must forecast cash flows and the rates at which various expenses and revenues will occur "if they are to maintain company liquidity and operating efficiiency" (Makridakis and Wheelwright, 1989).

Due to the nowadays difficult economic conditions that the whole business markets face up the importance of forecasting has become more imperative than ever. Marketing practitioners regard forecasting as an important part of their jobs. For example, in Dalrymple (1975), 93% of the companies sampled pointed out that sales forecasting was 'one of the most critical' aspects, or a 'very important' aspect of their company's success. Also Jobber, Hooley and Sanderson (1985), in a survey of 353 marketing directors from British textile firms, found that sales forecasting was the most common of nine activities on which they reported (Armostrong J. S. et. al, 2005).

Moreover Dalrymple (1987), in a survey among 134 US companies, found that 99% prepared formal forecasts when they developed written marketing plans.

Winklhofer et. al (1996) notes some basic factors that the importance of forecasting has become widely essential for the organizations in recent years:

The increasing complexity of organizations and their environments led to difficulties for decision makers to take account of all the factors relating to the future growth of the organization into account;

Organizations have moved towards more systematic decision making that contains explicit justifications for individual actions, and formalized forecasting is one way that these actions can be maintained;

The development of the forecasting methods has enables not only forecasting experts but also managers to become familiar with these techniques.

4

YIELD LOSS

Since the beginnings of agriculture about 10,000 years ago, farmers have had to compete with harmful organisms – pests. As with abiotic causes of crop losses, especially the lack or excess of water in the growth season, extreme temperatures, high or low irradiance and nutrient supply, biotic stressors have the poten tial to reduce crop production substantially. These organisms may be controlled by applying physical (cultivation, mechanical weeding etc.), biological (cultivar choice, crop rotation, antagonists, predators etc.) and chemical measures (pesticides).

Crop protection has been developed for the prevention and control of crop losses due to pests in the field. Pests reduce crop productivity in various ways, classified by their impacts; stand reducers (damping-off pathogens), photosynthetic rate reducers (fungi, bacteria, viruses), leaf senescence accelerators (pathogens), light stealers (weeds, some pathogens), assimilate sappers (nematodes, pathogens, sucking arthropods), and tissue consumers (chewing animals, necrotrophic pathogens).

Without preventive protection with pesticides, natural enemies, host plant resistance and other nonchemical controls, 70% of crops could have been lost to pests. Weeds produced the highest potential loss (30%), with animal pests and pathogens being less important (losses of 23 and 17%). The efficacy of control of pathogens and animal pests only reaches 32 and 39 %, respectively, compared to almost 74 % for weed control. Beside losses caused by pests, about 15% of losses are caused by economic damage and a further 15% because of harmful biological diversity.

The path to superior crop starts in the field, which means that every intervention in the field in greater or lesser extent will be reflected in the quality of the crop. Protection makes only 5-8% of the total cost during production in one vegetation. On the other hand, damages resulting from absence of proper care or poor performance can be huge. If all other interventions (fertilizing, weeding, tillage ...) are not performed properly, there will be a loss in the quality and quantity of farm yield. If protection is omitted or done superficially and inadequately, losses will be even higher.

Activities which maintain crops in terms of growth potential and also help us in the fight against pests are: fertilization, pruning, weeding, loosening the soil, irrigation, removal of crop residues from fields, crop rotation, and other works of tillage. These principles control a number of animal pests and diseases, some of which are certainly the most dangerous: downy mildew, powdery mildew, leaf rust, blight, thrips..

Good agricultural practices which unite the full range of preventive measures are extremely important. When planting permanent crops, you should choose the best location, which should be sunny and airy, where dew/humidity does not retain for long. Sow/plant varieties resistant to most diseases, regularly suppress weeds, keep an eye on conditions favorable to the development of diseases and pests. Using software tools which can help you fight pests and diseases is also a must.

The occurrence of most pests is caused by certain weather conditions; high humidity, low or high temperatures, rain, drought. Weather and temperature data are particularly helpful in following a pest's life cycle or in predicting how long it takes a certain pest to develop. Agrivi Farm Management Sofware knowledge base recognizes possible occurence of pests in your fields and alarms you 7 days in advance. Use this software to help you perform on time protection and reduce the risk of lower yield due to pests and diseases.

FACTORS IMPACTING UK WINTER WHEAT CROPS

Wheat is one of the main important cereal crops in the world. Also as one of the essential and important agricultural products. Wheat is considered as a strategic product due to its significant role in political and economic areas in the countries.

Loss of wheat may harm the economy in several aspects, including reduced production rates, increased need for imports and waste the inputs required for the production (Fallah and Rasouliazar, 2016).

In the UK average yields of winter wheat have increased progressively from the 1950s to the present a rate of approximately 100 kg ha-1 yr-1. Until the late 1980s, the increase was attributed about half to plant breeding and the half to husbandry. (Shearman et al., 2005) Any action was taken to improve the quantity and quality of the wheat would be of great importance.

Many factors came together for increasing yield of a winter wheat crop in the UK. Including plant establishment, rotation management, nutrient management, disease, pest and weed control, and soil management. Each factor can affect or limit yield performance. In this research, we investigate factors that influence winter wheat crops and which will improve yield performance.

Introduction

Cereal crops are a major staple food worldwide, contributing more than 50% of total human calorie input directly. In the last twenty years, the annual increment in the average yields of UK wheat and oilseed rape increased slightly. In the second half of the 20th-century cereal yields for example wheat, in the worldwide from 1 to 3 t haâ"1and in the UK approximately less than three t haâ"1 increased (Fischer and Edmeades, 2010; Hawkesford et al., 2013.)

Wheat is one of the important cereal crops in the world. Wheat is special in several ways for instance in human's food, they are used to make flour for yeast bread, or are blended with soft spring wheat to make the all- purpose flour used in a wide variety of baked products and The highest grain yields are obtained with winter wheat.

In this research, we investigate factors affecting winter wheat crops, and Evaluate of them can lead us to adjust management to achieve a greater yield.

Crop Nutrition

Each plant needs a different range of nutrients at every critical stage of its development. Cereal crops use nutrients for growth, progressively moving them from the roots and stems into the ear before the dying off and harvesting of the grain. For example, Phosphorus (P) is a vital component of adenosine triphosphate (ATP) which supplies the energy for many processes in the plant. Magnesium (Mg) is also a key component of leaf chlorophyll and the enzymes that support plant growth. Zinc and Boron directly influence the growth and yield of cereals and other crops, especially during flowering (Ghaffar et al., 2011).

Wheat as an important source of minerals such as iron, zinc, copper and magnesium. The dietary intake of these nutrients have decreased significantly in recent years in the UK which corresponds with the introduction of semi-dwarf, high-yielding cultivars. (Ming-Sheng Fan et al., 2008)

Semi-dwarf and high-yielding crop cultivars were responsible for increasing inputs of fertilisers and other agrochemicals which increased grain yield, however, increase in agronomic yield led to lower density of minerals in grain. The hypothesis of Davis et al indicated that decline in the nutrients in crops is because of changing in crops to varieties with improved yield. Ming-Sheng Fan et al, in the result of their experiments they also indicated that "decreasing mineral concentrations in wheat grain are partly due to a "dilution" effect resulting from increased yield.

It appears changing cultivars was affecting the mineral concentration. This change is further supported by (Flintham et al., 1997) that as a conclusion of the introduction of short-straw cultivars, HI (harvest index) increased, however mineral concentration significantly decreased in grain.

The solution for helping minerals in grain is to add micronutrients. Supplementation, diversification of diet and bio-fortification of crops by agronomic or genetic methods such as plant breeding can increase the micronutrients; the genetic methods are considered to be the most efficient for resource-poor populations in developing countries (Graham et al., 1999; Bouis, 2002; Welch and Graham, 2004). In the germplasm of the main crops there is a sufficient genetic variation in micronutrient concentrations, which can be explored in breeding strategies to combine the high nutrient density with the high-yielding traits(Graham et al.,1999; Cakmak et al ., 2004) such as Triticum Oflococum, Triticum dicoccol, and Triticum dicoccoides (Cakmak et al.,1999a, 2000; Ortiz-Monasterio and Graham ,2000)

Among wild wheat germplasm, the emmer wheat, Triticum dicoccoides, showed the largest variation and the highest concentration of micronutrients (Cakmak et al., 2000).

Crop rotation

One of the most effective management methods to prevent the development of the large population of plant diseases is crop rotation. Crop rotation means the planned order of specific crops planted on the same field (Huber et al., 1965), for example soil-borne pathogens which cause soil-borne diseases that lead to reducing crop yields can be controlled with rotations with non-host crops. Also, rotation suppresses Take all disease, Cephaelosporium stripe, Cercosporella foot rot and Cercosporella herpotrichoides (Welch et al., 2004). However, this method is used for preventing and it cannot reduce the size of the population significantly, therefore it is better to view crop rotation as preventive rather than curative.

Different crop rotations may have different effects on different pathogens, for instance Meloidogyne incognita were reduced by crops of crotalaria, marigold or bahiagrass while the population of Trichodorus christei were increased (Murphy et al.,1974), so it is vital to understand plant disease before planting crops.

- Crop Rotation benefits are = Maintains soil fertility.
- Reduces soil erosion.

- Controls pests, disease and weeds
- Reduces reliance on synthetic chemicals.

There are two major limits for crop rotation, the first one is some pathogens have many host ranges so it may be difficult to identify them, for example Rhizoctonia solani .The second one is that non-host crops may have little monetary value, so the selection of crops to use for rotation may be small. (Battese and Fuller, 1972) and if crops rotation not chosen properly, have chances of competition between the crops for nutrients.

Establishment management

One of the most important factors that can provide the correct condition for both growing and crop establishment is tillage. Tillage is accomplished to optimise productivity by alleviating physical, chemical and biological constraints of soil (Gajri et al., 2002).

Progressive tillage systems are developed to minimise cultivation costs and meet a range of variables such as soil or climate. It also improves the timeliness of planting which leads to an improved crop establishment. (Morris et al., 2010)

In the UK, current tillage systems can be divided into two general categories; Inversion tillage, known as conventional plough tillage, and Conservation tillage, known as non-inversion tillage.

Inversion tillage, inclusive of seed bed preparation by complete soil inversion, incorporates or buries most of the crop residue into the ground and consists of two processes: tillage and secondary tillage. Conservation tillage leaves the previous year's crop residue (such as wheat) on fields before and after planting the next crop using the two processes of strip tillage and direct drilling to reduce soil degradation and achieve a sustainable balance between production (Carter et al., 2003a) (Davies and Finney,2002).

No-till, known as direct drilling or zero tillage, is a conventional tillage system which sows directly in previous crops without any prior loosening of the soil cultivation. (Soane et al., 2012). Ploughing system is a farm implement used in farming for initial cultivation of soil in preparation for sowing seed or planting to loosen or turn the soil.

Ploughing system can expose soil compacted at harvest, therefore when it is loosened by weather it increases mixing of nutrients which is beneficial. Ploughing reduces risk of crop disease, pests and weeds. In contrast, no-till increases area capability and reduces overall costs such as fuel and machinery (Soane et al., 2012). Other limited and benefit of ploughing and no-till present in table 1.

Table 1. Agronomic advantages and disadvantages of No-till and ploughing

No till		Ploughing	
Advantages	Disadvantages	Advantages	Disadvantages
Reduce run-off, soil erosion and cost.	Not suitable for controling weeds, pest	Completely buries weeds and crop residue	High cost (Fuel and machinery)
Increase soil fertility	REduce grains yields	Increase mixing of nutrients when loosened by weather	Increased CO_2 emission
Increase area capabilityi	Crop residue remain on the soil surface	Reduced weeds crop diseases and pest	Extreme looseness to depth of ploughing.
Reduction of compaction below plough furrow	Crop establishment problems during very wet and very dry		High sensitivity to re-compaction of tops oil.

Pest management

Crop plants are attacked by many pests that affect plant survival, growth, and reproduction and as a result influence crop yield. Approximately 22% of the total area of wheat in the UK is affected by slugs without pesticides, Calculated that total annual cost to the UK industry from not controlling slugs in oilseed rape and wheat is approximately £43.5 million. (Nicholls, 2013).

Pests can cause crop-damaging by reducing rates of germination, feeding on seeds, roots, stem and leaves which reduce yields. In crop growth stages, pests can reduce yield potential of the harvest and Reduces crops ability to compensate for damage caused later in development and cause harm to crops near their harvest by reduces yield and quality. (Oerke, 2006; Damalas and Eleftherohorinos, 2011)

For example, two different aphid detriment in cereals crops in the UK because they transmit barley yellow dwarf luteovirus and Grain aphid Sitobion avenae. (Nicholls, 2013)

Despite these commonly acknowledged risks, pesticides do improve agricultural productivity. Pesticides can be considered as an economic, labor-saving, and efficient tool for pest management and can do improve agricultural productivity (Campbell,1976), such as Acetamiprid , Azadirachtin , Bifenthrin , Carbaryl for controlling insects but these benefits often go unnoticed by the general public.

Pesticides are widely used in the agricultural production hence, increasing amounts of pesticides must be applied at rising costs to maintain sufficient control. This increasing has led to integrated pest management (IPM) programs which reduce pest numbers to an acceptable threshold (Jain and Bhargava, 2007).

Push-pull strategies were conceived as a strategy for insect pest management (IPM) by Pyke et al. The push-pull technology is a strategy for controlling agricultural pests by using propulsive (push) while luring them toward an attractive source (pull) plants and trap them. This method can maximise the efficacy of behavior-manipulating stimuli through the additive and synergistic effects of integrating their use. Also can reduce pesticide input (Cook et al., 2006).

The other method that can effect on pests is no-till. It seems pest under no-till method decreases dramatically because of increased numbers of predators. For example, it reduces springtails (Onychiurus spp.) that cause damage in shoot and root of the sugar beet (Soane et al., 2012).However, the preservation of crop residues in wet conditions tends to increase slug population, and this can lead to damage to young seedlings in winter-sown Wheat and barley (Jordan et al., 1997).Using molluscicides may control slug population, but it will still increase production costs and affect beneficial soil biota (abid.)

SOIL MANAGEMENT

"Soil is the product of a complex set of interacting processes and cycles" (Fitzpatrick, 1991) Plants need a network of pore spaces to assistance gas exchange such as oxygen and carbon dioxide. Also water movement and nutrient uptake. Over time if the soil environment under intensive agricultural production was disturbed, the ability of the soil to maintain these conditions is compromised (Gerrard, 2000). So Preservation and improvement of soil quality are important in agricultural productivity and environmental quality.

Soil management practice which consideration all operations practices, and treatments used to protect the soil organic matter, soil structure, and the maintenance of a thriving soil microbial population. (Johnston and Fellow, 2005)

In central Great Plains, soil erosion and drought are the main problems associated with grain production. No- till is a good system for reducing soil erosion and maximise soil water conservation. (Fenster and Peterson, 1979)

In the UK using no-till after ploughing led to lower yields because of imperfect drainage and weak structure of soils during wet winter however all over the UK, good internal drainage was considered a pre-requisite for reliable success with no-till as discovered in the Germany (Ehlers and Claupein, 1994).

No-till, can also increase biological activity such as earthworms and improve stable root channels this can develop greater water permeability (Soane et al., 2012).

Under no-till, it seems, soil nutrients such as P and K increased dramatically near soil surface because of this microbiological activity. (Ehlers and Claupein, 1994). Also no-till it seems not suitable for sandy or drained soils. In Netherland, under the sandy soil, no-till decreased root penetration and yield such as root crop (Soane et al., 2012).

Weeds Management

Weeds have many attributes undesirable to crop producers and cause the adverse effect on their yield. There are limited reports on the inhibitory effects of weeds on crop plants (Bhowmik & Doll, 1992).Wheat (Triticum aestivum L.) is the most important cereal in the world. It has been estimated that globally yield reduction in wheat due to weeds is 13.1% (Oerke et al., 1994).

Weeds compete with the crop plants for sunlight, water, mineral nutrients and occupying a space, which would reduce the plant growth and this led to reducing crop yields, quality and harvesting efficiency. (Wright et al., 2001) Water requirement for the growth of weeds is primarily of interest from the standpoint of competition with the crop plant for the available moisture (Gibson, 2000).

Weeds provide a host for insects, nematodes and certain plant pathogens such as fungi. For instance Fusarium species pathogenic to winter wheat have been isolated from common broad-leaved weeds. (Jenkinson & Parry, 1994).

Weed management strategies attempt to eliminate or limit the deleterious effects of weeds when growing with crop plants and decrease weeds' ability to reduce yield. (Hager et al., 2003)

The basic of all Weed control methods is prevention. In few years the concept of prevention is not relevant anymore because of the availability of effective herbicides and mechanical control measures. These controlling tools have led to control weeds even after they have become established. (Walker 1995).

Herbicides may influence disease development via direct effects on the pathogens (Rodriguez-Kabana et al. 1966) or by altering plant resistance to pathogens (Dann et al. 1999; Levene et al. 1998) for example Broad-leaved weeds in winter wheat can be controlled by herbicides applied before or after wheat emergence in autumn or in spring. In Table 2 we mentioned a few Common herbicides used to control weeds

In recent years using Herbicides for controlling weed decreased because using herbicides may damage the crop. Also can get mixed with air, water and soil and kill other unwanted plant pests.So we must use other methods for controlling weeds without much dependence on chemicals like fertilisers and herbicides which have unpredictable harmful effects on the environment and human health. (Varshney et al., 2012)

Crop rotation is often identified as a valuable component of weed management for example Downy brome (Bromus tectorum L.) density remained relatively constant when winter wheat (Triticum aestivum L.) was rotated with oilseed rape (Blackshaw 1994a). (Abid.)

A cover crop is another method for managing weeds. They contribute organic matter to the soil, improve soil structure, and they can suppress weeds. Cover crops could permit reduction of herbicide inputs and a shift toward entire post-emergence herbicide programs for many crops. The early weed suppression provided by cover crop residue allow plants to become established before weed emergence. (Upadhyaya, and Blackshaw - 2007)

Table 2. Common herbicides used to control weeds.

Herbicide	Mode of action	Weeds controlled
Versatil	Absorbed by leaves, stems and roots.	Control thistles, yarrow, clovers and many difficult flat weeds.
Simazine	Absorbed only through roots of germinating plants	Prevents the emergence of a wide range of grasses and weeds
Buster	Systemic contact herbicide (via the leaf)	Broadleaved weeds and cloves, provides short-term weed control
Terbuthylazine	Absorbed through roots and leaves	Control a wide range of animal and perennial grasses and weeds

Wheat, it's an important food resource which approximately contains 40% of the world's population (Bockus, et al., 2010) annually global yield losses because of wheat diseases are estimated to be 20% in the field or the storage. (Abid.)

The disease can terminate entire trees in orchards or plantations Such as Phytophthora root or collar rot, and some can destroy before or after harvested product. For instance, the smuts, which destroy before or after the harvesting the cereal grains (all postharvest rots of fruits and vegetables) (James, 1974)

Each year in the UK the percentage of yield decreases because of disease diverse. The highest yield loss was between 1991, 1993 and 1998, and lowest was in 1995.Eyespot compared to other diseases caused greater yield losses. The second most significant disease was Powdery mildew in seven of the ten years but in 1990 demonstrate the greater losses than any other illness. The national economic impact of these yield losses varied depending on annual production figures and the price per tonne. (Hardwick et al., 2001) Each of these can infect wheat and cause disease throughout the growing season.

It is important to deliver consistently, high yields of high-quality grain thus controlling of cereal diseases is an important component of successful crop management. Recognition of the disease and an understanding of the pathogen's purpose are the first step for a successful disease control. Therefore disease management is necessary. Many strategies, tactics and techniques in disease management used in advanced (prevention) which applied before infection for protecting plant from disease and after infection controlled by heat or chemical treatment as bulbs, corms and fungicide (OC Maloy ,2005)

Plant disease management is a factor that has positive and negative externalities on the yield. The adverse effect includes environmental pollution; toxin remains in the soil, ecological damage, resource depletion, reduced disease management efficiency and costs associated with meeting minimum chemical residues on crop harvest. And positive effects can benefits to disease management in neighbouring farms, can reduce evolutionary potential of pathogens, and ensure social stability and safety (HE et al., 2016)

Farmers only pay the direct costs associated with plant disease management; they strongly select strategies that generate the best immediate economic returns while largely discounting potential negative impacts on the environment. To date, some highly effective disease management strategies have been used without sufficient regard to their long-term ecological effects such as Fungicide.

Fungicide was used on more than 93% of crops throughout the word for about ten years. In 1996 and 1998, fungicide usage reached 98%.it seems farmers do not value disease resistance but apply fungicides widely. Disease-induced losses are still substantial, despite the heavy reliance on fungicides (Stevens et al., 1997). During the 10-year using Fungicide has increased but so have septoria leaf blotch and eyespot. It has been suggested that national disease levels and yield losses were influenced more by season than fungicide usage. (Hardwick et al. 2000).

Priestley & Bayles (1980)found, from a postal survey of NIAB Fellows, that disease was seen more frequently in fungicide-treated (such as benomyl, carbendazim) than in untreated fields. Cook & King (1984) indicated that final disease levels were often similar in treated and untreated crops.

For improving yield, it would be better for plant disease management to focus on increasing crop productivity, using fewer fungicides, reducing food contamination by microbial toxins, Change the philosophy of plant disease management to focus on pathogens (or insect vectors) to managing host plants for achieving high productivity yield and Considerations in agricultural productivity and crop health. (Problems, challenges and future of plant disease management: from an ecological point of view)

EFFECT OF AZADIRACHTIN ON INSECTS

Humans have always been in direct competition with a myriad of insects, pests from our ancestral beginning. This competition for food with insects intensified when humans began to cultivate plants converting the natural ecosystem to an agroecosystem. Also insects serve as vectors of various diseases caused by bacterial, filarial nematode, protozoans and viruses. Therefore control of insects posed a major concern for the development of the economy. In 1939, the discovery of insecticidal properties of DDT (Dichlorodiphenyl trichloroethane) by Paul H. Mueller changed the scenario of pest management. During World War 2 DDT was extensively used to prevent epidemics of several insect vectored diseases such as yellow fever, typhus elephantiasis and malaria. This drew attention to the possibilities of more synthetic insecticides and as a result the use of pesticides in various arenas soared from 1940-60, complete reliance on pesticides intensive pest management was leading agriculture on a "pesticide treadmill". The overreliance on synthetic pesticides from late 1940s to mid-1960 was referred to as "Dark Ages" of pest control. The cheapness and effectiveness of synthetic insecticides threw natural compounds into shade. But very soon other shades also began to appear. In 1962, the appearance of book "Silent Spring" (by Rachel Carison) showed that pesticide residues were building up in ecosystem with detrimental effects on wildlife and beneficial insects. Due to extensive and intensive use, misuse and abuse of insecticides the following problems were becoming prominent and intolerable (ecological backlashes):

1. Development of insecticide resistance- many insects started developing resistance against pesticides which increased the cost of management.
2. Due to killing of natural controlling agents, the phenomenon of pest resurgence became more evident.
3. Also minor pest were achieving the status of major pest i.e. secondary pest outbreak due to significant decline in its natural enemy (predators and parasites).
4. Ecological imbalance due to poisoning of all the realms of environment.
5. Increase in the concentration of hydrochlorinated insecticide in food chain.

6. Intolerable residues on the food made the food obtained after such treatment uneconomical as it became unfit for consumption and unfit for exports due to high toxic residues.
7. Killing and harmful effects on Non target organisms became more prominent (like birds, fishes and other wildlife).

Overviewing these effects, there was an utmost need for the development of environmentally sound management practices. This lead to the idea of Integrated Pest Management (IPM) . A panel of experts put the concept of IPM in 1968. IPM as defined by FAO is a system which in consideration with the present environment and pest population dynamics, integrates all the sustainable techniques of pest management as compatible a manner as possible and maintain the population of pest below the level which can cause economic damage (i.e. below economic injury level). The approach is to minimize the dependence on insecticides and maximize the use of ecofriendly methods so as to cause minimum damage to the environment. Botanical pesticides, thus is an very important component of IPM as

1. They are easily degradable.
2. Don't affect non target organisms, natural controlling agents such as predator, parasites.
3. Don't form residues
4. And has no harmful effect on humans as they are very specific in action.

Botanical pesticides refer to the use of chemical or organic compound produced by plants, plant products, which have harmful effects on the growth, development and survival of insect pests. Plants are a rich source of such organic compounds.

HISTORY

The practice of utilizing the derivatives of plant i.e. botanical pesticides in agriculture dates at least two millennia back in ancient China, Egypt, Greece and India. Even in North America and Europe, the documented use of botanicals extend back more than 150 years before the discovery of major class of synthetic chemical insecticides (OP, carbamates and pyretheroids) in mid 1930s to 1950. It is very clear from the recent history that the chemical insecticides have essentially relegated the botanical pesticides from an important role in agriculture to a trivial position in the market among various crop protection strategies.

The total number of 20 phylochemicals is estimated to be 500000, so far only 10000 of these have been isolated. At present four major types of botanicals are being used for the control of insects. These include:

a. Pyrethrum

b. Neem (Azadirachtin).

c. Rotenons.

d. Essential oils.

Others are in limited use like Ryania, Nicotine, Sabdella. Whereas Nicotine, Rotenene, Natural Pyrethrins constitute the outstanding example of older botanicals, extracts and compounds from the Neem tree (A.indica) have emerged as the most prominent phytochemical pesticides in recent years. Among the various biologically active compounds that can be extracted from the Neem tree like- triterpenoid, phenolic compounds, carotenoids, steroids, ketones; the tetranortriterpenoid azadirachtin has been the most extensively studied pesticide as 1) it is relatively abundant in Neem kernels. 2) has biological activity on a wide range of insects.

PLANT SECONDARY METABOLITES

Plants produce a large, diverse array of organic compounds that appears to have no function in growth and development. These substances are known as SECONDARY METABOLITES or secondary products or natural products. Secondary metabolites differ from Primary metabolites (amino acids, nucleotides, sugars, acyl lipids) as:

a. They have no direct roles in photosynthesis, respiration, protein synthesis etc

b. They have restricted distribution in plant kingdom.

In a seminal paper Fraenkel stressed the role of secondary metabolites as defense system against insects, pests and other natural enemies.

Though they play no role in growth and metabolism they play important ecological role in plants:

a. They protect plants against being eaten by herbivore and being infected by microbial pathogen.

b. They serve as attractants for pollinators and seed dispersing animals and as agents of plant-plant competition.

Because of their ecological role, plant secondary metabolites are classified as ALLELOCHEMICALS, a term coined by WHITTAKER. An allelochmical is defined as a non nutritional chemical produced by an individual of one species that affects growth, health, behavior, population ecology of another species. Plants produce an astonishing array of Secondary metabolites. Even a single plant species may produce an extensive pharmacopeia of recondite chemicals. Periwinkle for example contains about more than 100 monoterpenoid indole alkaloids. It has been estimated that plant kingdom synthesizes hundreds of thousands of different secondary metabolites. The no of identified compounds now exceeds 10000.

Plant secondary metabolites can be divided into three chemically distinct groups:

A. Terpenes
B. Phenolics
C. Nitrogen Containing Compound

TERPENES

The terpenes constitute the largest class of secondary products .the diverse substances of this class are generally insoluble in water. They are biosynthesized from acetyl coA.Terpenes are classified by no of five carbon units they contain as:

1. Monoterpenes: Contain 2 five carbon skeleton
2. Sesquiterpenes: Contain 3 five carbon skeleton
3. Diterpenes: Contain 4 five carbon skeleton
4. Triterpenes: 30 carbons
5. Tetraterpenes: 40 carbons
6. Polyterpenoids: $(C5)n$,where $n>8$

Some terpenes have role in growth and development

Terpenes defend against herbivore in many plants. Terpenes are toxins and feeding deterrentsto many plant feeding insects, thus they appear to play important defensive role in plant kingdom and protection of agricultural crops. Examples of important Terpenes:

1. **PYRETHROIDS:** These are monoterpenoid that occurs in leaves and flowers of Chrysanthemum species show very striking insecticidal activity. Both natural and synthetic pyrethroids are popular ingredients in commercial insecticide because of their low persistence in the environment. Pyrethrum is the predominant botanical in use accounting for 80% of global botanical insecticide.

2. **ESSENTIAL OILS:** These are the mixture of monoterpene and sesquiterpene that lends a characteristics odor to the foliage .e.g Menttholin

Peppermint oil and Limonenein lemon oil are monoterpenes. Essential oils have well known insect repellent properties. They are frequently found in glandular hairs and serve to advertize the toxicity of plant repelling potential. Phytophagus insects even they take a trial bite.

3. **VOLATILE TERPENES:** In corn & wild tobacco certain monoterpenes and sesquiterpenes are produced and emitted only after insect feeding has already begun. These substances prevent oviposition and kill plant feeding insects and so help in controlling further damage .These also attract natural enemies of plant feeding insects so promise a sound means of pest control.

4. **LIMNOIDS:** These are a group of nonvolatile Triterpene.Among these the most powerful deterrent to insects feeding known is Azadirachtin.It is a complex limnoid from Neem tree which is feeding deterrent to some insects at as low as 50ppm and it exerts a variety of toxic effect .It has considerable potential as a commercial insect control because of its low toxicity to mammals.

PHENOLICS

Plants produce a variety of secondary products that contain a phenol group, these are called phenolic compounds. Plants phenolics are a chemically heterogeneous group of nearly 10000 compounds .many of these serve as defense compounds against herbivores.

- The release of phenolics into soil limits the growth of other plants.
- LIGNIN a highly branched polymer of phenylpropanoid group has significant protective function in plants. Its physical toughness deters feeding by insects and chemical durability makes it relatively indigestible.
- The flavoids are one of the largest classes of plants phenolics e.g.anthocyanins, flavones etc.
- Anthocyanins are colored flavonoids that attract insects to flower and fruits by providing visual and olfactory signal.
- Flavonoids protect against damage by UV light.
- Tannins deter feeding by herbivores and it also act as feeding repellents to a great diversity of insects

NITROGEN CONTAINING COMPOUND

A large variety of plant secondary metabolites have nitrogen in their structure. This category includes well known defense against phytophagus insects as alkaloids and cyanogenic glycosides,glucosinolates.

ALKALOIDS: These are a large family of more than 15000 nitrogen containing secondary metabolites with a heterocyclic ring. Several different types including nicotine and its relative are derived from ornithine .Most alkaloids now function as defenseagainst their predators because of their toxicity and deterrence capability. Alkaloids increase in response to initial damage fortifying against further damage e.g. wild tobacco produces higher level of nicotinefollowing damage by tobacco caterpillars.

CYANOGENIC GLYCOSIDES: These are not toxic themselves but are readily broken down to give off volatile poisons; well known poisonous gas Hydrogen cyanide. When the leaf is damaged due to insects feeding on it, the cell content of different tissue mix and HCN is formed .HCN is a fast acting toxin that inhibits metalloprotiens such as iron containing cytochrome oxidase; a key enzyme of mitochondrial respiration, thus affecting physiology of insects. Thus presence of cyanogenic glycosides deters feeding by insects.

GLUCOSINOLATES: A Class of plant glycosides that break down to release volatile defensive substances, also called Mustard oil glycosides.Found principally in the Brassicaceae and related plant families, where glucosinolates give off compounds responsible for smell and taste of vegetables like cabbage, cauliflower, mustards etc. These compounds function in DEFENCE as toxin and feeding repellent. But certain insects are adapted for feeding on glucosinolate containing plants without ill effects .For example glucosinolates serve as stimulant for Cabbage butterfly for feeding and egg laying and isothiocyanates serve as volatile attractants.

PLANT PROTIENS: Certain plant protein also interfere with insect digestion, for example plants produce LECTINS,defensive proteins that bind to epithelial cell lining digestive tract and interfere with nutrient absorption .The best known anti digestive proteins in plants are protein inhibitors found in legumes, tomatoes, and other plants. After entering herbivores digestive tract they interfere with protein digestion, as a result insects suffer reduced rates of growth and development.

The neem tree, properties and distribution

The Neem tree also known by names like Indian Lilac, Margosa tree is an evergreen fastgrowing tree belonging to the order "Rutales" and family "Meliaceae". The genus Azadirachta indica was described by A.juss in 1830.

Disribution

Neem tree is indigenous to Indian Subcontinent from where is has spread to many Asian and African countries such as Pakistan, Bangladesh, Mynamar, Sri Lanka, Thailand, Indonesia, Malaysia, Singapore, Iran, Yemen, Australia, New

Guinea, Nigeria, Fizi, Tanzania, Madagascar, USA, Latin America, Germany, France, Portugal, Spain and UK. It is now grown in most tropical and sub-tropical parts of the worls.

The origin of A.indica is not very clear. Some say that is has originated from Burma whereas others point it to south India. It is considered that it has originated from south-eastern and southern Asia.

In Indonesia Neem exists in low lying Northern and Eastern parts of java. In Philippines it was introduced from India, Africa. Ketkar (1967) reported about 14 million trees in India. There are more than 20 million trees available in entire India. In Africa Neem was introduced from India and is concentrated in a belt stretching across the African continent from Somalia to Mauretania. In America Neem trees are prominent in Haiti, Surinam and propagation has started in Brazil, Puerto Rico, Cuba and Nicaragua. Neem trees also grow in our neighbouring countries, Middle East, Saudi Arabia and Yemen.

Habitat

Neem tree is a fast growing sclerophyllous tree.

- It grows well in humid to semi-humid climate.
- It thrives well at altitudes upto 700-800m above the sea level.
- Neem trees are hardy and are able to grow in severe drought condition also. They thrive well in regions with less than 500 mm annual rainfall and upto 2500 mm annual rainfall.
- Neem tree exist in poor, shallow, sandy and stony soil. It also grows in black cotton soil in India.
- Neem tree can flourish in warm to very hot climates. It grows well between 21-320c temperatures but it can tolerate upto 500c during summer.
- Ph value between 6.2-7 seems to the best for the growth of Neem tree.

CHARACTERISTIC FEATURES

- It is a fast-growing tree, reaching a height of 4-7 m during the first 3 years and 5-11m during the following 5 years.
- It begins to bear fruit within 3-5 years and becomes fully productive in the 10th year, when it may yield up to 50 kg fruit per tree per year.
- The Neem tree produces its fruits, which are the main source for its production of pesticides, on drooping panicles, usually about once a year,

although two fruiting periods per year occur in certain areas (e.g. West Africa).

- A mature Neem tree produces annually 30-50 kg of fruit, but this may depend upon rainfall and soil conditions. More conservative estimates range around 20 kg per tree; 40 kg of fresh fruit yield about 24 kg of dry fruit.
- Neem has the reputation of possessing a large number of biological activities which include insecticidal, nematicidal, bactericidal, and anti-fungal. It has attracted world-wide attention due to its wide ranging capacity as a biocide.

PESTICIDAL CONSTITUENT OF NEEM

Neem tree is the only tree in which every part of tree produces biologically active products which has various properties such as antifeedant, deterrent, growth regulation, oviposition alteration, insecticidal properties, fungicidal properties,etc.

Though bark, heartwood, leaves, fruits of it produce these substances in various concentrations but it is the fruits specifically seeds which are of major importance. Neem seed kernels contain the highest amount of the active compound. 40-50 kg of fruit can yield about 5 kg of kernels (10% of fruit). Each seed contains about 1-3 kernels.

Till date more than 140 active principles have been identified in different parts of the tree. Insecticidal properties of Neem is due to the presence of a class of Limnoids which include compounds like Azadirachtin, Melantriol, Salanin, Mimbines, Salannol and various sulfur containing compounds. Among these Azadirachtin is the most active and predominant insecticidal compound concentrated mainly in the seed kernels. The Azadirachtin occurs in seeds at the concentration of about 0.1-0.9%. It is estimated that 20-30 kg of Neem seeds are required per hectare if 2g of Azadirachtin per kg of sed is obtained. The highest yield of Azadirachtin obtained till date was about 10g/kg of seed.

5

PESTICIDE AND INSECTICIDES

According to the Environmental Protection Agency (EPA), the government body that regulates pesticides in the U.S., a pesticide is any substance or mixture of substances intended for preventing, destroying, repelling or mitigating any pest. Though often misunderstood to refer only to insecticides, the term pesticide also applies to herbicides, fungicides, and various other substances used to control pests. Pesticides also include plant regulators, defoliants and desiccants.

WHAT'S IN A PESTICIDE PRODUCT?

We normally think of a pesticide as the product that can be purchased in the store – the insecticide, the weed killer or the fungicide. But, unfortunately, there is much more to it than that. The product that you buy or are exposed to is actually a pesticide formulation that contains a number of different materials, including active and inert ingredients, as well as contaminants and impurities. In addition, pesticides, when subject to various environmental conditions, break down to other materials known as metabolites, which are sometimes more toxic than the parent material.

Active Ingredients

The active ingredient, usually the only component of the formulation listed on the pesticide label, is by nature biologically and chemically active against a target pest, be it an insect, weed or fungus. By definition these chemicals kill living things.

Contaminants and Impurities

Contaminants and impurities are often a part of the pesticide product and responsible for product hazards. Dioxin and DDT have been identified as contaminants, which have not been purposefully added but are a function of the production process.

Metabolites

Metabolites are breakdown products that form when a pesticide is used in the environment and mixes with air, water, soil or living organisms. Often the metabolite is more hazardous than the parent pesticide.

Inert Ingredients

If you were to go to your local hardware store and take a look at the label on a can of ant and roach killer, the contents might read something like this, "5% Permethrin, 95% Inert Ingredients." After reading the label, you may wonder what makes up the other 95%. The fact is, the manufacturer doesn't have to tell you. Currently, under the Federal Insecticide, Fungicide and Rodenticide Act (FIFRA), pesticide manufacturers are only required to list the active ingredients in a pesticide, leaving consumers and applicators unaware of the possible toxics present in the inert ingredients of pesticide products they are using, unless the EPA administrator determines that the chemical poses a public health threat. Pesticide manufacturers argue they cannot release information on inert ingredients because they are trade secrets, and if released, their products could be duplicated. Quite often inert ingredients constitute over 95% of the pesticide product. Inert ingredients are mixed into pesticides products as a carrier or sticking agent, and are often as toxic as the active ingredient.

The Hazards of Inert Ingredients

Despite their name, these ingredients are neither chemically, biologically or toxicologically inert. In general, inert ingredients are minimally tested, however, many are known to state, federal and international agencies to be hazardous to human health. For example, the U.S. government lists creosols as a "Hazardous Waste" under Superfund regulations, yet allows these chemicals to be listed as inert ingredients in pesticide products. Creosols are known to produce skin and eye irritations, burns, inflammation, blindness, pneumonia, pancreatitis, central nervous system depression and kidney failure.

Some inert ingredients are even more toxic than the active ingredients. One of the most hazardous ingredients in the commonly used herbicide RoundUp is a surfactant, which is classified as an inert, and therefore not listed on the label. The pesticide naphthalene is an inert ingredient in some products and listed as an active ingredient in others. According a 2000 report produced by the New York State Attorney General, The Secret Ingredients in Pesticides: Reducing the Risk, 72 percent of pesticide products available to consumers contain over 95 percent inert ingredients; fewer than 10 percent of pesticide products list any inert ingredients on their labels; more than 200 chemicals used as inert ingredients are hazardous pollutants in federal environmental statutes governing air and water quality; and, of a 1995 list of inert ingredients, 394 chemicals were listed as active ingredients in other pesticide products.

Definition

The Food and Agriculture Organization (FAO) has defined pesticide as: any substance or mixture of substances intended for preventing, destroying, or controlling any pest, including vectors of human or animal disease, unwanted species of plants or animals, causing harm during or otherwise interfering with the production, processing, storage, transport, or marketing of food, agricultural commodities, wood and wood products or animal feedstuffs, or substances that may be administered to animals for the control of insects, arachnids, or other pests in or on their bodies. The term includes substances intended for use as a plant growth regulator, defoliant, desiccant, or agent for thinning fruit or preventing the premature fall of fruit. Also used as substances applied to crops either before or after harvest to protect the commodity from deterioration during storage and transport.

Pesticides can be classified by target organism (e.g., herbicides, insecticides, fungicides, rodenticides, and pediculicides – see table), chemical structure (e.g., organic, inorganic, synthetic, or biological (biopesticide), although the distinction can sometimes blur), and physical state (e.g. gaseous (fumigant)). Biopesticides include microbial pesticides and biochemical pesticides. Plant-derived pesticides, or "botanicals", have been developing quickly. These include the pyrethroids, rotenoids, nicotinoids, and a fourth group that includes strychnine and scilliroside.

Many pesticides can be grouped into chemical families. Prominent insecticide families include organochlorines, organophosphates, and carbamates. Organochlorine hydrocarbons (e.g., DDT) could be separated into dichlorodiphenylethanes, cyclodiene compounds, and other related compounds. They operate by disrupting the sodium/potassium balance of the nerve fiber, forcing the nerve to transmit continuously. Their toxicities vary greatly, but they have been phased out because of their persistence and potential to bioaccumulate. Organophosphate and carbamates largely replaced organochlorines. Both operate through inhibiting the enzyme acetylcholinesterase, allowing acetylcholine to transfer nerve impulses indefinitely and causing a variety of symptoms such as weakness or paralysis. Organophosphates are quite toxic to vertebrates and have in some cases been replaced by less toxic carbamates. Thiocarbamate and dithiocarbamates are subclasses of carbamates. Prominent families of herbicides include phenoxy and benzoic acid herbicides (e.g. 2,4-D), triazines (e.g., atrazine), ureas (e.g., diuron), and Chloroacetanilides (e.g., alachlor). Phenoxy compounds tend to selectively kill broad-leaf weeds rather than grasses. The phenoxy and benzoic acid herbicides function similar to plant growth hormones, and grow cells without normal cell division, crushing the plant's nutrient transport system. Triazines interfere with photosynthesis. Many commonly used pesticides are not included in these families, including glyphosate.

Pesticides can be classified based upon their biological mechanism function or application method. Most pesticides work by poisoning pests. A systemic pesticide moves inside a plant following absorption by the plant. With insecticides and most fungicides, this movement is usually upward (through the xylem) and outward. Increased efficiency may be a result. Systemic insecticides, which poison pollen and nectar in the flowers, may kill bees and other needed pollinators.

In 2009, the development of a new class of fungicides called paldoxins was announced. These work by taking advantage of natural defense chemicals released by plants called phytoalexins, which fungi then detoxify using enzymes. The paldoxins inhibit the fungi's detoxification enzymes. They are believed to be safer and greener.

Uses

Pesticides are used to control organisms that are considered to be harmful. For example, they are used to kill mosquitoes that can transmit potentially deadly diseases like West Nile virus, yellow fever, and malaria. They can also kill bees, wasps or ants that can cause allergic reactions. Insecticides can protect animals from illnesses that can be caused by parasites such as fleas. Pesticides can prevent sickness in humans that could be caused by moldy food or diseased produce. Herbicides can be used to clear roadside weeds, trees and brush. They can also kill invasive weeds that may cause environmental damage. Herbicides are commonly applied in ponds and lakes to control algae and plants such as water grasses that can interfere with activities like swimming and fishing and cause the water to look or smell unpleasant. Uncontrolled pests such as termites and mold can damage structures such as houses. Pesticides are used in grocery stores and food storage facilities to manage rodents and insects that infest food such as grain. Each use of a pesticide carries some associated risk. Proper pesticide use decreases these associated risks to a level deemed acceptable by pesticide regulatory agencies such as the United States Environmental Protection Agency (EPA) and the Pest Management Regulatory Agency (PMRA) of Canada.

DDT, sprayed on the walls of houses, is an organochlorine that has been used to fight malaria since the 1950s. Recent policy statements by the World Health

Organization have given stronger support to this approach. However, DDT and other organochlorine pesticides have been banned in most countries worldwide because of their persistence in the environment and human toxicity. DDT use is not always effective, as resistance to DDT was identified in Africa as early as 1955, and by 1972 nineteen species of mosquito worldwide were resistant to DDT.

Amount used

In 2006 and 2007, the world used approximately 2.4 megatonnes (5.3×10 lb) of pesticides, with herbicides constituting the biggest part of the world pesticide use at 40%, followed by insecticides (17%) and fungicides (10%). In 2006 and 2007 the U.S. used approximately 0.5 megatonnes (1.1×10 lb) of pesticides, accounting for 22% of the world total, including 857 million pounds (389 kt) of conventional pesticides, which are used in the agricultural sector (80% of conventional pesticide use) as well as the industrial, commercial, governmental and home & garden sectors. Pesticides are also found in majority of U.S. households with 78 million out of the 105.5 million households indicating that they use some form of pesticide. As of 2007, there were more than 1,055 active ingredients registered as pesticides, which yield over 20,000 pesticide products that are marketed in the United States.

The US used some 1 kg (2.2 pounds) per hectare of arable land compared with: 4.7 kg in China, 1.3 kg in the UK, 0.1 kg in Cameroon, 5.9 kg in Japan and 2.5 kg in Italy. Insecticide use in the US has declined by more than half since 1980, (.6%/yr) mostly due to the near phase-out of organophosphates. In corn fields, the decline was even steeper, due to the switchover to transgenic Bt corn.

For the global market of crop protection products, market analysts forecast revenues of over 52 billion US$ in 2019.

Benefits

Pesticides can save farmers' money by preventing crop losses to insects and other pests; in the U.S., farmers get an estimated fourfold return on money they spend on pesticides.One study found that not using pesticides reduced crop yields by about 10%. Another study, conducted in 1999, found that a ban on pesticides in the United States may result in a rise of food prices, loss of jobs, and an increase in world hunger.

There are two levels of benefits for pesticide use, primary and secondary. Primary benefits are direct gains from the use of pesticides and secondary benefits are effects that are more long-term.

Primary benefits

Controlling pests and plant disease vectors

- Improved crop/livestock yields
- Improved crop/livestock quality
- Invasive species controlled

Controlling human/livestock disease vectors and nuisance organisms

- Human lives saved and suffering reduced
- Animal lives saved and suffering reduced
- Diseases contained geographically

Controlling organisms that harm other human activities and structures

- Drivers view unobstructed
- Tree/brush/leaf hazards prevented
- Wooden structures protected

Monetary

Every dollar ($1) that is spent on pesticides for crops yields four dollars ($4) in crops saved. This means based that, on the amount of money spent per year on pesticides, $10 billion, there is an additional $40 billion savings in crop that would be lost due to damage by insects and weeds. In general, farmers benefit from having an increase in crop yield and from being able to grow a variety of crops throughout the year. Consumers of agricultural products also benefit from being able to afford the vast quantities of produce available year-round. The general public also benefits from the use of pesticides for the control of insect-borne diseases and illnesses, such as malaria. The use of pesticides creates a large job market within the agrichemical sector.

Health effects of pesticides

Health effects of pesticides may be acute or delayed in those who are exposed. A 2007 systematic review found that "most studies on non-Hodgkin lymphoma and leukemia showed positive associations with pesticide exposure" and thus concluded that cosmetic use of pesticides should be decreased. Strong evidence also exists for other negative outcomes from pesticide exposure including neurological problems, birth defects, fetal death, and neurodevelopmental disorder.

According to The Stockholm Convention on Persistent Organic Pollutants, 9 of the 12 most dangerous and persistent chemicals are pesticides.

ACUTE EFFECTS

Acute health problems may occur in workers that handle pesticides, such as abdominal pain, dizziness, headaches, nausea, vomiting, as well as skin and eye problems. In China, an estimated half million people are poisoned by pesticides each year, 500 of whom die. Pyrethrins, insecticides commonly used in common bug killers, can cause a potentially deadly condition if breathed in.

Long-term effects

Cancer

Many studies have examined the effects of pesticide exposure on the risk of cancer. Associations have been found with: leukemia, lymphoma, brain, kidney, breast, prostate, pancreas, liver, lung, and skin cancers. This increased risk occurs with both residential and occupational exposures. Increased rates of cancer have been found among farm workers who apply these chemicals. A mother's occupational exposure to pesticides during pregnancy is associated with an increases in her child's risk of leukemia, Wilms' tumor, and brain cancer. Exposure to insecticides within the home and herbicides outside is associated with blood cancers in children.

Neurological

Evidence links pesticide exposure to worsened neurological outcomes.

The United States Environmental Protection Agency finished a 10-year review of the organophosphate pesticides following the 1996 Food Quality Protection Act, but did little to account for developmental neurotoxic effects, drawing strong criticism from within the agency and from outside researchers. Comparable studies have not been done with newer pesticides that are replacing organophosphates.

Reproductive effects

Strong evidence links pesticide exposure to birth defects, fetal death and altered fetal growth. Agent Orange, a 50:50 mixture of 2,4,5-T and 2,4-D, has been associated with bad health and genetic effects in Malaya and Vietnam. It was also found that offspring that were at some point exposed to pesticides had a low birth weight and had developmental defects.

Fertility

A number of pesticides including dibromochlorophane and 2,4-D has been associated with impaired fertility in males. Pesticide exposure resulted in reduced fertility in males, genetic alterations in sperm, a reduced number of sperm, damage to germinal epithelium and altered hormone function.

Other

Some studies have found increased risks of dermatitis in those exposed.

Additionally, studies have indicated that pesticide exposure is associated with long-term respiratory problems. Summaries of peer-reviewed research have examined the link between pesticide exposure and neurologic outcomes and cancer, perhaps the two most significant things resulting in organophosphate-exposed workers.

According to researchers from the National Institutes of Health (NIH), licensed pesticide applicators who used chlorinated pesticides on more than 100 days in their lifetime were at greater risk of diabetes. One study found that associations between specific pesticides and incident diabetes ranged from a 20 percent to a 200 percent increase in risk. New cases of diabetes were reported by 3.4 percent of those in the lowest pesticide use category compared with 4.6 percent of those in the highest category. Risks were greater when users of specific pesticides were compared with applicators who never applied that chemical.

Route of exposure

People can be exposed to pesticides by a number of different routes including: occupation, in the home, at school and in their food.

There are concerns that pesticides used to control pests on food crops are dangerous to people who consume those foods. These concerns are one reason for the organic foodmovement. Many food crops, including fruits and vegetables, contain pesticide residues after being washed or peeled. Chemicals that are no longer used but that are resistant to breakdown for long periods may remain in soil and water and thus in food.

The United Nations Codex Alimentarius Commission has recommended international standards for maximum residue limits (MRLs), for individual pesticides in food.

In the EU, MRLs are set by DG-SANCO.

In the United States, levels of residues that remain on foods are limited to tolerance levels that are established by the U.S. Environmental Protection Agency and are considered safe. The EPA sets the tolerances based on the toxicity of the pesticide and its breakdown products, the amount and frequency of pesticide application, and how much of the pesticide (i.e., the residue) remains in or on food by the time it is marketed and prepared. Tolerance levels are obtained using scientific risk assessments that pesticide manufacturers are required to produce by conducting toxicological studies, exposure modeling and residue studies before a particular pesticide can be registered, however, the effects are tested for single pesticides, and there is little information on possible synergistic effects of exposure to multiple pesticide traces in the air, food and water.

Strawberries and tomatoes are the two crops with the most intensive use of soil fumigants. They are particularly vulnerable to several type of diseases, insects, mites, and parasitic worms. In 2003, in California alone, 3.7 million pounds (1,700 metric tons) of metham sodium were used on tomatoes. In recent years other farmers have demonstrated that it is possible to produce strawberries and tomatoes without the use of harmful chemicals and in a cost-effective way.

Exposure routes other than consuming food that contains residues, in particular pesticide drift, are potentially significant to the general public.

Some pesticides can remain in the environment for prolonged periods of time. For example, most people in the United States still have detectable levels of DDT in their bodies even though it was banned in the US in 1972.

PREVENTION

Pesticides exposure cannot be studied in placebo controlled trials as this would be unethical. A definitive cause effect relationship therefore cannot be established. Consistent evidence can and has been gathered through other study designs. The precautionary principle is thus frequently used in environmental law such that absolute proof is not required before efforts to decrease exposure to potential toxins are enacted.

The American Medical Association recommend limiting exposure to pesticides. They came to this conclusion due to the fact that surveillance systems currently in place are inadequate to determine problems related to exposure. The utility of applicator certification and public notification programs are also of unknown value in their ability to prevent adverse outcomes.

Epidemiology

The World Health Organization and the UN Environment Programme estimate that each year, 3 million workers in agriculture in the developing world experience severe poisoning from pesticides, about 18,000 of whom die. According to one study, as many as 25 million workers in developing countries may suffer mild pesticide poisoning yearly.Detectable levels of 50 different pesticides were found in the blood of a representative sample of the U.S. population.

Society and culture

Concerns regarding conflict of interests regarding the research base have been raised. A number of researchers involved with pesticides have been found to have undisclosed ties to industry including: Richard Doll or the Imperial Cancer Research Fund in England and Hans-Olov Adami of the Karolinska Institute in Sweden.

Other animals

A number of pesticides including clothianidin, dinotefuran, imidacloprid are toxic to bees. Exposure to pesticides may be one of the contributory factors to colony collapse disorder. A study in North Carolina indicated that more than 30 percent of the quail tested were made sick by one aerial insecticide application. Once sick, wild birds may neglect their young, abandon their nests, and become more susceptible to predators or disease.

PESTICIDE POISONING

A pesticide poisoning occurs when chemicals intended to control a pest affect non-target organisms such as humans, wildlife, or bees. There are three types of pesticide poisoning. The first of the three is a single and short-term very high level of exposure which can be experienced by individuals who commit suicide, as well as pesticide formulators. The second type of poisoning is long-term high-level exposure, which can occur in pesticide formulators and manufacturers. The third type of poisoning is a long-term low-level exposure, which individuals are exposed to from sources such as pesticide residues in food as well as contact with pesticide residues in the air, water, soil, sediment, food materials, plants and animals.

In developing countries, such as Sri Lanka, pesticide poisonings from short-term very high level of exposure (acute poisoning) is the most worrisome type of poisoning. However, in developed countries, such as Canada, it is the complete opposite: acute pesticide poisoning is controlled, thus making the main issue long-term low-level exposure of pesticides.

Cause

The most common exposure scenarios for pesticide-poisoning cases are accidental or suicidal poisonings, occupational exposure, by-stander exposure to off-target drift, and the general public who are exposed through environmental contamination.

Accidental or suicidal?

Self-poisoning with agricultural pesticides represents a major hidden public health problem accounting for approximately one-third of all suicides worldwide. It is one of the most common forms of self-injury in the Global South. The World Health Organization estimates that 300,000 people die from self-harm each year in the Asia-Pacific region alone. Most cases of intentional pesticide poisoning appear to be impulsive acts undertaken during stressful events, and the availability of pesticides strongly influences the incidence of self poisoning. Pesticides are the agents most frequently used by farmers and students in India to commit suicide.

Occupational

Pesticide poisoning is an important occupational health issue because pesticides are used in a large number of industries, which puts many different categories of workers at risk. Extensive use puts agricultural workers in particular at increased risk for pesticide illnesses. Workers in other industries are at risk for exposure as well. For example, commercial availability of pesticides in stores puts retail workers at risk for exposure and illness when they handle pesticide products. The ubiquity of pesticides puts emergency responders such as fire-fighters and police officers at risk, because they are often the first responders to emergency events and may be unaware of the presence of a poisoning hazard. The process of aircraft disinsection, in which pesticides are used on inbound international flights for insect and disease control, can also make flight attendants sick.

Different job functions can lead to different levels of exposure. Most occupational exposures are caused by absorption through exposed skin such as the face, hands, forearms, neck, and chest. This exposure is sometimes enhanced by inhalation in settings including spraying operations in greenhouses and other closed environments, tractor cabs, and the operation of rotary fan mist sprayers.

RESIDENTIAL

When thinking of pesticide poisoning, one does not take into consideration the contribution that is made of their own household. The majority of households in Canada use pesticides while taking part in activities such as gardening. In Canada 96 percent of households report having a lawn or a garden. 56 percent of the households who have a lawn or a garden utilize fertilizer or pesticide. This form of pesticide use may contribute to the third type of poisoning, which is caused by long-term low-level exposure. As mentioned before, long-term low-level exposure affects individuals from sources such as pesticide residues in food as well as contact with pesticide residues in the air, water, soil, sediment, food materials, plants and animals.

Pathophysiology

Organochlorines

The organochlorine pesticides, like DDT, aldrin, and dieldrin are extremely persistent and accumulate in fatty tissue. Through the process of bioaccumulation (lower amounts in the environment get magnified sequentially up the food chain), large amounts of organochlorines can accumulate in top species like humans. There is substantial evidence to suggest that DDT, and its metabolite DDE, act as endocrine disruptors, interfering with hormonal function of estrogen, testosterone, and other steroid hormones.

Anticholinesterase compounds

Certain organophosphates have long been known to cause a delayed-onset toxicity to nerve cells, which is often irreversible. Several studies have shown persistent deficits in cognitive function in workers chronically exposed to pesticides. Newer evidence suggests that these pesticides may cause developmental neurotoxicity at much lower doses and without depression of plasma cholinesteraselevels.

Diagnosis

Most pesticide-related illnesses have signs and symptoms that are similar to common medical conditions, so a complete and detailed environmental and occupational history is essential for correctly diagnosing a pesticide poisoning. A few additional screening questions about the patient's work and home environment, in addition to a typical health questionnaire, can indicate whether there was a potential pesticide poisoning.

If one is regularly using carbamate and organophosphate pesticides, it is important to obtain a baseline cholinesterase test. Cholinesterase is an important enzyme of the nervous system, and these chemical groups kill pests and potentially injure or kill humans by inhibiting cholinesterase. If one has had a baseline test and later suspects a poisoning, one can identify the extent of the problem by comparison of the current cholinesterase level with the baseline level.

Prevention

Accidental poisonings can be avoided by proper labeling and storage of containers. When handling or applying pesticides, exposure can be significantly reduced by protecting certain parts of the body where the skin shows increased absorption, such as the scrotal region, underarms, face, scalp, and hands. Using chemical-resistant gloves has been shown to reduce contamination by 33–86%.

Further methods in order to aid prevention of acute pesticide poisoning, concerning both accidental death and suicides, there could be a method for national governments to control accessibility. The pesticides most toxic to humans if restricted has the possibility to reduce deaths. There could also be designated locations in rural living areas and cities used to safely store toxic pesticides in order to gain control over usage.

Treatment

Specific treatments for acute pesticide poisoning are often dependent on the pesticide or class of pesticide responsible for the poisoning. However, there are basic management techniques that are applicable to most acute poisonings, including skin decontamination, airway protection, gastrointestinal decontamination, and seizure treatment. Decontamination of the skin is performed while other life-saving measures are taking place. Clothing is removed, the patient is showered with soap and water, and the hair is shampooed to remove chemicals from the skin and hair. The eyes are flushed with water for 10–15 minutes. The patient is intubated and oxygen administered, if necessary. In more severe cases, pulmonary ventilation must sometimes be supported mechanically. Seizures are typically managed with lorazepam, phenytoin and phenobarbitol, or diazepam(particularly for organochlorine poisonings).

Gastric lavage is not recommended to be used routinely in pesticide poisoning management, as clinical benefit has not been confirmed in controlled studies; it is indicated only when the patient has ingested a potentially life-threatening amount of poison and presents within 60 minutes of ingestion. An orogastric tube is inserted and the stomach is flushed with saline to try to remove the poison. If the patient is neurologically impaired, a cuffed endotracheal tube inserted beforehand for airway protection. Studies of poison recovery at 60 minutes have shown recovery of 8%–32%. However, there is also evidence that lavage may flush the material into the small intestine, increasing absorption. Lavage is contra-indicated in cases of hydrocarbon ingestion.

Activated charcoal is sometimes administered as it has been shown to be successful with some pesticides. Studies have shown that it can reduce the amount absorbed if given within 60 minutes, though there is not enough data to determine if it is effective if time from ingestion is prolonged. Syrup of ipecac is no longer recommended for most pesticide poisonings.

Urinary alkalinisation has been used in acute poisonings from chlorophenoxy herbicides (such as 2,4-D, MCPA, 2,4,5-T and mecoprop); however, evidence to support its use is poor.

EPIDEMIOLOGY

Acute pesticide poisoning is a large-scale problem, especially in developing countries.

"Most estimates concerning the extent of acute pesticide poisoning have been based on data from hospital admissions which would include only the more serious cases. The latest estimate by a WHO task group indicates that there may be 1 million serious unintentional poisonings each year and in addition 2 million people hospitalized for suicide attempts with pesticides. This necessarily reflects only a fraction of the real problem. On the basis of a survey of self-reported minor poisoning carried out in the Asian region, it is estimated that there could be as many as 25 million agricultural workers in the developing world suffering an episode of poisoning each year." In Canada in 2007 more than 6000 cases of acute pesticide poisoning occurred.

Estimating the numbers of chronic poisonings worldwide is more difficult.

Society and culture

Rachel Carson's Silent Spring brought about the first major wave of public concern over the chronic effects of pesticides.

In other animals

An obvious side effect of using a chemical meant to kill is that one is likely to kill more than just the desired organism. Contact with a sprayed plant or "weed" can have an effect upon local wildlife, most notably insects. A cause for concern is how pests, the reason for pesticide use, are building up a resistance. Phytophagous insects are able to build up this resistance because they are easily capable of evolutionary diversification and adaptation. The problem this presents is that in order to obtain the same desired effect of the pesticides they have to be made increasingly stronger as time goes on. Repercussions of the use of stronger pesticides on vegetation has a negative result on the surrounding environment, but also would contribute to consumers' long-term low-level exposure.

Environmental impact of pesticides

The impact of pesticides consists of the effects of pesticides on non-target species. Pesticides are chemical preparations used to kill fungal or animal pests. Over 98% of sprayed insecticides and 95% of herbicides reach a destination other than their target species, because they are sprayed or spread across entire agricultural fields. Runoff can carry pesticides into aquatic environments while wind can carry them to other fields, grazing areas, human settlements and undeveloped areas, potentially affecting other species. Other problems emerge from poor production, transport and storage practices. Over time, repeated application increases pest resistance, while its effects on other species can facilitate the pest's resurgence.

Each pesticide or pesticide class comes with a specific set of environmental concerns. Such undesirable effects have led many pesticides to be banned, while regulations have limited and/or reduced the use of others. Over time, pesticides have generally become less persistent and more species-specific, reducing their environmental footprint. In addition the amounts of pesticides applied per hectare have declined, in some cases by 99%. However, the global spread of pesticide use, including the use of older/obsolete pesticides that have been banned in some jurisdictions, has increased overall.

Agriculture and the environment

The arrival of humans in an area, to live or to conduct agriculture, necessarily has environmental impacts. These range from simple crowding out of wild plants in favor of more desirable cultivars to larger scale impacts such as reducing biodiversity by reducing food availability of native species, which can propagate across food chains. The use of agricultural chemicals such as fertilizer and pesticides magnify those impacts. While advances in agrochemistry have reduced those impacts, for example by the replacement of long-lived chemicals with those that reliably degrade, even in the best case they remain substantial. These effects are magnified by the use of older chemistries and poor management practices.

Air

Pesticides can contribute to air pollution. Pesticide drift occurs when pesticides suspended in the air as particles are carried by wind to other areas, potentially contaminating them. Pesticides that are applied to crops can volatilize and may be blown by winds into nearby areas, potentially posing a threat to wildlife. Weather conditions at the time of application as well as temperature and relative humidity change the spread of the pesticide in the air. As wind velocity increases so does the spray drift and exposure. Low relative humidity and high temperature result in more spray evaporating. The amount of inhalable pesticides in the outdoor environment is therefore often dependent on the season. Also, droplets of sprayed pesticides or particles from pesticides applied as dusts may travel on the wind to other areas, or pesticides may adhere to particles that blow in the wind, such as dust particles. Ground spraying produces less pesticide drift than aerial spraying does. Farmers can employ a buffer zone around their crop, consisting of empty land or non-crop plants such as evergreen trees to serve as windbreaks and absorb the pesticides, preventing drift into other areas. Such windbreaks are legally required in the Netherlands. Pesticides that are sprayed on to fields and used to fumigate soil can give off chemicals called volatile organic compounds, which can react with other chemicals and form a pollutant called tropospheric ozone. Pesticide use accounts for about 6 percent of total tropospheric ozone levels.

Water

In the United States, pesticides were found to pollute every stream and over 90% of wells sampled in a study by the US Geological Survey. Pesticide residues have also been found in rain and groundwater. Studies by the UK government showed that pesticide concentrations exceeded those allowable for drinking water in some samples of river water and groundwater.

Pesticide impacts on aquatic systems are often studied using a hydrology transport model to study movement and fate of chemicals in rivers and streams. As early as the 1970s quantitative analysis of pesticide runoff was conducted in order to predict amounts of pesticide that would reach surface waters.

There are four major routes through which pesticides reach the water: it may drift outside of the intended area when it is sprayed, it may percolate, or leach, through the soil, it may be carried to the water as runoff, or it may be spilled, for example accidentally or through neglect. They may also be carried to water by eroding soil. Factors that affect a pesticide's ability to contaminate water include its water solubility, the distance from an application site to a body of water, weather, soil type, presence of a growing crop, and the method used to apply the chemical.

UNITED STATES REGULATIONS

In the US, maximum limits of allowable concentrations for individual pesticides in drinking water are set by the Environmental Protection Agency (EPA) for public water systems. (There are no federal standards for private wells.) Ambient water quality standards for pesticide concentrations in water bodies are principally developed by state environmental agencies, with EPA oversight. These standards may be issued for individual water bodies, or may apply statewide.

United Kingdom regulations

The United Kingdom sets Environmental Quality Standards (EQS), or maximum allowable concentrations of some pesticides in bodies of water above which toxicity may occur.

European Union regulations

The European Union also regulates maximum concentrations of pesticides in water.

Soil

Many of the chemicals used in pesticides are persistent soil contaminants, whose impact may endure for decades and adversely affect soil conservation.

The use of pesticides decreases the general biodiversity in the soil. Not using the chemicals results in higher soil quality, with the additional effect that more organic matter in the soil allows for higher water retention. This helps increase yields for farms in drought years, when organic farms have had yields 20-40% higher than their conventional counterparts. A smaller content of organic matter in the soil increases the amount of pesticide that will leave the area of application, because organic matter binds to and helps break down pesticides.

Degradation and sorption are both factors which influence the persistence of pesticides in soil. Depending on the chemical nature of the pesticide, such processes control directly the transportation from soil to water, and in turn to air and our food. Breaking down organic substances, degradation, involves interactions among microorganisms in the soil. Sorption affects bioaccumulation of pesticides which are dependent on organic matter in the soil. Weak organic acids have been shown to be weakly sorbed by soil, because of pH and mostly acidic structure. Sorbed chemicals have been shown to be less accessible to microorganisms. Aging mechanisms are poorly understood but as residence times in soil increase, pesticide residues become more resistant to degradation and extraction as they lose biological activity.

EFFECT ON PLANTS

Nitrogen fixation, which is required for the growth of higher plants, is hindered by pesticides in soil. The insecticides DDT, methyl parathion, and especially pentachlorophenol have been shown to interfere with legume-rhizobium chemical signaling. Reduction of this symbiotic chemical signaling results in reduced nitrogen fixation and thus reduced crop yields. Root nodule formation in these plants saves the world economy $10 billion in synthetic nitrogen fertilizer every year.

Pesticides can kill bees and are strongly implicated in pollinator decline, the loss of species that pollinate plants, including through the mechanism of Colony Collapse Disorder, in which worker bees from a beehive or western honey bee colony abruptly disappear. Application of pesticides to crops that are in bloom can kill honeybees, which act as pollinators. The USDA and USFWS estimate that US farmers lose at least $200 million a year from reduced crop pollination because pesticides applied to fields eliminate about a fifth of honeybee colonies in the US and harm an additional 15%.

On the other side, pesticides have some direct harmful effect on plant including poor root hair development, shoot yellowing and reduced plant growth.

Effect on animals

Many kinds of animals are harmed by pesticides, leading many countries to regulate pesticide usage through Biodiversity Action Plans.

Animals including humans may be poisoned by pesticide residues that remain on food, for example when wild animals enter sprayed fields or nearby areas shortly after spraying.

Pesticides can eliminate some animals' essential food sources, causing the animals to relocate, change their diet or starve. Residues can travel up the food chain; for example, birds can be harmed when they eat insects and worms that have consumed pesticides. Earthworms digest organic matter and increase nutrient content in the top layer of soil. They protect human health by ingesting decomposing litter and serving as bioindicators of soil activity. Pesticides have had harmful effects on growth and reproduction on earthworms.Some pesticides can bioaccumulate, or build up to toxic levels in the bodies of organisms that consume them over time, a phenomenon that impacts species high on the food chain especially hard.

Birds

The US Fish and Wildlife Service estimates that 72 million birds are killed by pesticides in the United States each year. Bald eagles are common examples of nontarget organisms that are impacted by pesticide use. Rachel Carson's book Silent Spring dealt with damage to bird species due to pesticide bioaccumulation. There is evidence that birds are continuing to be harmed by pesticide use. In the farmland of the United Kingdom, populations of ten different bird species declined by 10 million breeding individuals between 1979 and 1999, allegedly from loss of plant and invertebrate species on which the birds feed. Throughout Europe, 116 species of birds were threatened as of 1999. Reductions in bird populations have been found to be associated with times and areas in which pesticides are used. DDE-induced egg shell thinning has especially affected European and North American bird populations. In another example, some types of fungicides used in peanut farming are only slightly toxic to birds and mammals, but may kill earthworms, which can in turn reduce populations of the birds and mammals that feed on them.

Some pesticides come in granular form. Wildlife may eat the granules, mistaking them for grains of food. A few granules of a pesticide may be enough to kill a small bird.

The herbicide paraquat, when sprayed onto bird eggs, causes growth abnormalities in embryos and reduces the number of chicks that hatch successfully, but most herbicides do not directly cause much harm to birds. Herbicides may endanger bird populations by reducing their habitat.

Aquatic life

Fish and other aquatic biota may be harmed by pesticide-contaminated water. Pesticide surface runoff into rivers and streams can be highly lethal to aquatic life, sometimes killing all the fish in a particular stream.

Application of herbicides to bodies of water can cause fish kills when the dead plants decay and consume the water's oxygen, suffocating the fish. Herbicides such as copper sulfite that are applied to water to kill plants are toxic to fish and other water animals at concentrations similar to those used to kill the plants. Repeated exposure to sublethal doses of some pesticides can cause physiological and behavioral changes that reduce fish populations, such as abandonment of nests and broods, decreased immunity to disease and decreased predator avoidance.

Application of herbicides to bodies of water can kill plants on which fish depend for their habitat.

Pesticides can accumulate in bodies of water to levels that kill off zooplankton, the main source of food for young fish. Pesticides can also kill off insects on which some fish feed, causing the fish to travel farther in search of food and exposing them to greater risk from predators.

The faster a given pesticide breaks down in the environment, the less threat it poses to aquatic life. Insecticides are typically more toxic to aquatic life than herbicides and fungicides.

Amphibians

In the past several decades, amphibian populations have declined across the world, for unexplained reasons which are thought to be varied but of which pesticides may be a part.

Pesticide mixtures appear to have a cumulative toxic effect on frogs. Tadpoles from ponds containing multiple pesticides take longer to metamorphose and are smaller when they do, decreasing their ability to catch prey and avoid predators. Exposing tadpoles to the organochloride endosulfan at levels likely to be found in habitats near fields sprayed with the chemical kills the tadpoles and causes behavioral and growth abnormalities.

The herbicide atrazine can turn male frogs into hermaphrodites, decreasing their ability to reproduce. Both reproductive and nonreproductive effects in aquatic reptiles and amphibians have been reported. Crocodiles, many turtle species and some lizards lack sex-distinct chromosomes until after fertilization during organogenesis, depending on temperature. Embryonic exposure in turtles to various PCBs causes a sex reversal. Across the United States and Canada disorders such as decreased hatching success, feminization, skin lesions, and other developmental abnormalities have been reported.

PEST REBOUND AND SECONDARY PEST OUTBREAKS

Non-target organisms can also be impacted by pesticides. In some cases, a pest insect that is controlled by a beneficial predator or parasite can flourish should an insecticide application kill both pest and beneficial populations. A study comparing biological pest control and pyrethroid insecticide for diamondback moths, a major cabbage family insect pest, showed that the pest population rebounded due to loss of insect predators, whereas the biocontrol did not show the same effect. Likewise, pesticides sprayed to control mosquitoes may temporarily depress mosquito populations, however they may result in a larger population in the long run by damaging natural controls. This phenomenon, wherein the population of a pest species rebounds to equal or greater numbers than it had before pesticide use, is called pest resurgence and can be linked to elimination of its predators and other natural enemies.

Loss of predator species can also lead to a related phenomenon called secondary pest outbreaks, an increase in problems from species that were not originally a problem due to loss of their predators or parasites. An estimated third of the 300 most damaging insects in the US were originally secondary pests and only became a major problem after the use of pesticides. In both pest resurgence and secondary outbreaks, their natural enemies were more susceptible to the pesticides than the pests themselves, in some cases causing the pest population to be higher than it was before the use of pesticide.

Eliminating pesticides

Many alternatives are available to reduce the effects pesticides have on the environment. Alternatives include manual removal, applying heat, covering weeds with plastic, placing traps and lures, removing pest breeding sites, maintaining healthy soils that breed healthy, more resistant plants, cropping native species that are naturally more resistant to native pests and supporting biocontrol agents such as birds and other pest predators. In the United States, conventional pesticide use peaked in 1979, and by 2007, had been reduced by 25 percent from the 1979 peak level, while US agricultural output increased by 43 percent over the same period.

Biological controls such as resistant plant varieties and the use of pheromones, have been successful and at times permanently resolve a pest problem. Integrated Pest Management (IPM) employs chemical use only when other alternatives are ineffective. IPM causes less harm to humans and the environment. The focus is broader than on a specific pest, considering a range of pest control alternatives. Biotechnology can also be an innovative way to control pests. Strains can be genetically modified (GM) to increase their resistance to pests. However the same techniques can be used to increase pesticide resistance and was employed by Monsanto to create glyphosate-resistant strains of major crops. In 2010, 70% of all the corn that was planted was resistant to glyphosate; 78% of cotton, and 93% of all soybeans.

Alternatives

Alternatives to pesticides are available and include methods of cultivation, use of biological pest controls (such as pheromones and microbial pesticides), genetic engineering, and methods of interfering with insect breeding. Application of composted yard waste has also been used as a way of controlling pests. These methods are becoming increasingly popular and often are safer than traditional chemical pesticides. In addition, EPA is registering reduced-risk conventional pesticides in increasing numbers.

Cultivation practices include polyculture (growing multiple types of plants), crop rotation, planting crops in areas where the pests that damage them do not live, timing planting according to when pests will be least problematic, and use of trap crops that attract pests away from the real crop. Trap crops have successfully controlled pests in some commercial agricultural systems while reducing pesticide usage; however, in many other systems, trap crops can fail to reduce pest densities at a commercial scale, even when the trap crop works in controlled experiments. In the U.S., farmers have had success controlling insects by spraying with hot water at a cost that is about the same as pesticide spraying.

Release of other organisms that fight the pest is another example of an alternative to pesticide use. These organisms can include natural predators or parasites of the pests. Biological pesticides based on entomopathogenic fungi, bacteria and viruses cause disease in the pest species can also be used.

Interfering with insects' reproduction can be accomplished by sterilizing males of the target species and releasing them, so that they mate with females but do not produce offspring. This technique was first used on the screwworm fly in 1958 and has since been used with the medfly, the tsetse fly, and the gypsy moth. However, this can be a costly, time consuming approach that only works on some types of insects. Agroecology emphasize nutrient recycling, use of locally available and renewable resources, adaptation to local conditions, utilization of microenvironments, reliance on indigenous knowledge and yield maximization while maintaining soil productivity. Agroecology also emphasizes empowering people and local communities to contribute to development, and encouraging "multi-directional" communications rather than the conventional "top-down" method.

PUSH–PULL AGRICULTURAL PEST MANAGEMENT

The push–pull technology is a strategy for controlling agricultural pests by using repellent "push" plants and trap "pull" plants. For example, cereal crops like maize or sorghumare often infested by stem borers. Grasses planted around the perimeter of the crop attract and trap the pests, whereas other plants, like Desmodium, planted between the rows of maize repel the pests and control the parasitic plant Striga. Push–pull technology was developed at the International Centre of Insect Physiology and Ecology (ICIPE) in Kenya in collaboration with Rothamsted Research, UK and national partners.

The pull

The approach relies on a combination of companion crops to be planted around and among maize or sorghum. Both domestic and wild grasses can help to protect the crops by attracting and trapping the stemborers. The grasses are planted in the border around the maize and sorghum fields where invading adult moths become attracted to chemicals emitted by the grasses themselves. Instead of landing on the maize or sorghum plants, the insects head for what appears to be a tastier meal. These grasses provide the "pull" in the "push–pull" strategy. They also serve as a haven for the borers' natural enemies. Good trap crops include well-known grasses such as Napier grass (Pennisetum purpureum) and Sudan grass (Sorghum vulgare sudanense). Napier grass has a particularly effective way of defending itself against the pests: once attacked by a borer larva, it secretes a sticky substance which physically traps the pest and limits its damage.

The push

The "push" in the intercropping scheme is provided by the plants that emit chemicals (kairomones) which repel stemborer moths and drive them away from the main crop (maize or sorghum). The best candidates discovered so far with the repellent properties are species of leguminous genus Desmodium. Desmodium is planted in between the rows of maize or sorghum. Being a low-growing plant, it does not interfere with the crops' growth and, furthermore, has the advantage of maintaining soil stability, improving soil fertility through enhanced soil organic matter content and nitrogen fixation. It also serves as a highly nutritious animal feed and effectively suppresses striga weeds. Another plant showing good repellent properties is molasses grass (Melinis minutiflora), a nutritious animal feed with tick-repelling and stemborer larval parasitoid attractive properties.

HOW PUSH–PULL WORKS

The push–pull technology involves use of behaviour-modifying stimuli to manipulate the distribution and abundance of stemborers and beneficial insects for management of stemborer pests. It is based on in-depth understanding of chemical ecology, agrobiodiversity, plant-plant and insect-plant interactions, and involves intercropping a cereal crop with a repellent intercrop such as Desmodium uncinatum (silverleaf) (push), with an attractive trap plant such as Napier grass (pull) planted as a border crop around this intercrop. Gravid stemborer females are repelled from the main crop and are simultaneously attracted to the trap crop. Napier grass produces significantly higher levels of attractive volatile compounds (green leaf volatiles), cues used by gravid stemborer females to locate host plants, than maize or sorghum. There is also an increase of approximately 100-fold in the total amounts of these compounds produced in the first hour of nightfall by Napier grass (scotophase), the period at which stemborer moths seek host plants for laying eggs, causing the differential oviposition preference. However, many of the stemborer larvae, about 80%, do not survive, as Napier grass tissues produce sticky sap in response to feeding by the larvae, which traps them, causing their mortality. Legumes in the Desmodium genus (silverleaf, D. uncinatum and greenleaf, D. intortum), on the other hand, produce repellent volatile chemicals that push away the stemborer moths. These include (E)-â-ocimene and (E)-4,8-dimethyl-1,3,7-

nonatriene, semiochemicals produced during damage to plants by herbivorous insects and are responsible for the repellence of Desmodium to stemborers.

Desmodium also controls striga, resulting in significant yield increases of about 2 tonnes/hectare (0.9 short tons per acre) per cropping season. In the elucidation of the mechanisms of striga suppression by D. uncinatum, it was found that, in addition to benefits derived from increased availability of nitrogen and soil shading, an allelopathic effect of the root exudates of the legume, produced independently of the presence of striga, is responsible for this dramatic reduction in an intercrop with maize. Presence of blends of secondary metabolites with striga seed germination stimulatory, 42 2 ,52 2 ,-dihydro-5,22 ,42 -trihydroxy-52 2 ,-isopropenylfurano-(22 2 ,32 2 ;7,6)-isoflavanone, and postgermination inhibitory, 42 2 ,52 2 -dihydro-22 -methoxy-5,42 -dihydroxy-52 2 -isopropenylfurano- (22 2 ,32 2 ;7,6)-isoflavanone, activities in the root exudates of D. uncinatum which directly interferes with parasitism was observed. This combination thus provides a novel means of in situ reduction of the striga seed bank in the soil through efficient suicidal germination even in the presence of grassy host plants in the proximity. Other Desmodium species have also been evaluated and have similar effects on stemborers and striga weed and are currently being used as intercrops in maize, sorghum and millets.

Types

Pesticides are often referred to according to the type of pest they control. Pesticides can also be considered as either biodegradable pesticides, which will be broken down by microbes and other living beings into harmless compounds, or persistent pesticides, which may take months or years before they are broken down: it was the persistence of DDT, for example, which led to its accumulation in the food chain and its killing of birds of prey at the top of the food chain. Another way to think about pesticides is to consider those that are chemical pesticides are derived from a common source or production method.

Some examples of chemically-related pesticides are:

Neonicotinoid pesticides

Neonicotinoids are a class of neuro-active insecticides chemically similar to nicotine. Imidacloprid, of the neonicotanoid family, is the most widely used insecticide in the world. In the late 1990s neonicotinoids came under increasing scrutiny over their environmental impact and were linked in a range of studies to adverse ecological effects, including honey-beecolony collapse disorder (CCD) and loss of birds due to a reduction in insect populations. In 2013, the European Union and a few non EU countries restricted the use of certain neonicotinoids.

Organophosphate pesticides

Organophosphates affect the nervous system by disrupting acetylcholinesterase activity, the enzyme that regulates acetylcholine, a neurotransmitter. Most organophosphates are insecticides. They were developed during the early 19th century, but their effects on insects, which are similar to their effects on humans, were discovered in 1932.Some are very poisonous. However, they usually are not persistent in the environment.

Carbamate pesticides

Carbamate pesticides affect the nervous system by disrupting an enzyme that regulates acetylcholine, a neurotransmitter. The enzyme effects are usually reversible. There are several subgroups within the carbamates.

Organochlorine insecticides

They were commonly used in the past, but many have been removed from the market due to their health and environmental effects and their persistence (e.g., DDT, chlordane, and toxaphene).

PYRETHROID PESTICIDES

They were developed as a synthetic version of the naturally occurring pesticide pyrethrin, which is found in chrysanthemums. They have been modified to increase their stability in the environment. Some synthetic pyrethroids are toxic to the nervous system.

Sulfonylurea herbicides

The following sulfonylureas have been commercialized for weed control: amidosulfuron, azimsulfuron, bensulfuron-methyl, chlorimuron-ethyl, ethoxysulfuron, flazasulfuron, flupyrsulfuron-methyl-sodium, halosulfuron-methyl, imazosulfuron, nicosulfuron, oxasulfuron, primisulfuron-methyl, pyrazosulfuron-ethyl, rimsulfuron, sulfometuron-methyl Sulfosulfuron, terbacil, bispyribac-sodium, cyclosulfamuron, and pyrithiobac-sodium. Nicosulfuron, triflusulfuron methyl, and chlorsulfuron are broad-spectrum herbicides that kill plants weeds or pests by inhibiting the enzyme acetolactate synthase. In the 1960s, more than 1 kg/ha (0.89 lb/acre) crop protection chemical was typically applied, while sulfonylureates allow as little as 1% as much material to achieve the same effect.

Biopesticides

Biopesticides are certain types of pesticides derived from such natural materials as animals, plants, bacteria, and certain minerals. For example, canola oil and baking soda have pesticidal applications and are considered biopesticides. Biopesticides fall into three major classes:

- Microbial pesticides which consist of bacteria, entomopathogenic fungi or viruses (and sometimes includes the metabolites that bacteria or fungi produce). Entomopathogenic nematodes are also often classed as microbial pesticides, even though they are multi-cellular.
- Biochemical pesticides or herbal pesticides are naturally occurring substances that control (or monitor in the case of pheromones) pests and microbial diseases.
- Plant-incorporated protectants (PIPs) have genetic material from other species incorporated into their genetic material (i.e. GM crops). Their use is controversial, especially in many European countries.

Further types of pesticides

- The term pesticide also include these substances:
- Defoliants : Cause leaves or other foliage to drop from a plant, usually to facilitate harvest.
- Desiccants : Promote drying of living tissues, such as unwanted plant tops.
- Insect growth regulators : Disrupt the molting, maturity from pupal stage to adult, or other life processes of insects.
- Plant growth regulators : Substances (excluding fertilizers or other plant nutrients) that alter the expected growth, flowering, or reproduction rate of plants.
- Wood preservatives : They are used to make wood resistant to insects, fungus, and other pests.

Regulation

International

In most countries, pesticides must be approved for sale and use by a government agency.

In Europe, recent EU legislation has been approved banning the use of highly toxic pesticides including those that are carcinogenic, mutagenic or toxic to reproduction, those that are endocrine-disrupting, and those that are persistent, bioaccumulative and toxic (PBT) or very persistent and very bioaccumulative (vPvB). Measures were approved to improve the general safety of pesticides across all EU member states.

Though pesticide regulations differ from country to country, pesticides, and products on which they were used are traded across international borders. To deal with inconsistencies in regulations among countries, delegates to a conference of the United Nations Food and Agriculture Organization adopted an International Code of Conduct on the Distribution and Use of Pesticides in 1985 to create voluntary standards of pesticide regulation for different countries. The Code was updated in 1998 and 2002. The FAO claims that the code has raised awareness about pesticide hazards and decreased the number of countries without restrictions on pesticide use.

Three other efforts to improve regulation of international pesticide trade are the United Nations London Guidelines for the Exchange of Information on Chemicals in International Trade and the United Nations Codex Alimentarius Commission. The former seeks to implement procedures for ensuring that prior informed consent exists between countries buying and selling pesticides, while the latter seeks to create uniform standards for maximum levels of pesticide residues among participating countries. Both initiatives operate on a voluntary basis.

Pesticides safety education and pesticide applicator regulation are designed to protect the public from pesticide misuse, but do not eliminate all misuse. Reducing the use of pesticides and choosing less toxic pesticides may reduce risks placed on society and the environment from pesticide use. Integrated pest management, the use of multiple approaches to control pests, is becoming widespread and has been used with success in countries such as Indonesia, China, Bangladesh, the U.S., Australia, and Mexico. IPM attempts to recognize the more widespread impacts of an action on an ecosystem, so that natural balances are not upset. New pesticides are being developed, including biological and botanical derivatives and alternatives that are thought to reduce health and environmental risks. In addition, applicators are being encouraged to consider alternative controls and adopt methods that reduce the use of chemical pesticides.

Pesticides can be created that are targeted to a specific pest's lifecycle, which can be environmentally more friendly. For example, potato cyst nematodes emerge from their protective cysts in response to a chemical excreted by potatoes; they feed on the potatoes and damage the crop. A similar chemical can be applied to fields early, before the potatoes are planted, causing the nematodes to emerge early and starve in the absence of potatoes.

UNITED STATES

In the United States, the Environmental Protection Agency (EPA) is responsible for regulating pesticides under the Federal Insecticide, Fungicide, and Rodenticide Act (FIFRA) and the Food Quality Protection Act (FQPA). Studies must be conducted to establish the conditions in which the material is safe to use and the effectiveness against the intended pest(s). The EPA regulates pesticides to ensure that these products do not pose adverse effects to humans or the environment. Pesticides produced before November 1984 continue to be reassessed in order to meet the current scientific and regulatory standards. All registered pesticides are reviewed every 15 years to ensure they meet the proper standards. During the registration process, a label is created. The label contains directions for proper use of the material in addition to safety restrictions. Based on acute toxicity, pesticides are assigned to a Toxicity Class.

Some pesticides are considered too hazardous for sale to the general public and are designated restricted use pesticides. Only certified applicators, who have passed an exam, may purchase or supervise the application of restricted use pesticides. Records of sales and use are required to be maintained and may be audited by government agencies charged with the enforcement of pesticide regulations.These records must be made available to employees and state or territorial environmental regulatory agencies.

The EPA regulates pesticides under two main acts, both of which amended by the Food Quality Protection Act of 1996. In addition to the EPA, the United States Department of Agriculture (USDA) and the United States Food and Drug Administration (FDA) set standards for the level of pesticide residue that is allowed on or in crops. The EPA looks at what the potential human health and environmental effects might be associated with the use of the pesticide.

In addition, the U.S. EPA uses the National Research Council's four-step process for human health risk assessment: (1) Hazard Identification, (2) Dose-Response Assessment, (3) Exposure Assessment, and (4) Risk Characterization. Recently Kaua'i County (Hawai'i) passed Bill No. 2491 to add an article to Chapter 22 of the county's code relating to pesticides and GMOs. The bill strengthens protections of local communities in Kaua'i where many large pesticide companies test their products.

Insecticides

Insecticides are pesticides that are formulated to kill, harm, repel or mitigate one or more species of insect. Insecticides work in different ways. Some insecticides disrupt the nervous system, whereas others may damage their exoskeletons, repel them or control them by some other means. They can also be packaged in various forms including sprays, dusts, gels, and baits. Because of these factors, each insecticide can pose a different level of risk to non-target insects, people, pets and the environment.

An insecticide is a substance used to kill insects. They include ovicides and larvicides used against insect eggs and larvae, respectively. Insecticides are used in agriculture, medicine, industry and by consumers. Insecticides are claimed to be a major factor behind the increase in the 20th century's agricultural productivity. Nearly all insecticides have the potential to significantly alter ecosystems; many are toxic to humans; some concentrate along the food chain.

Insecticides can be classified in two major groups: systemic insecticides, which have residual or long term activity; and contact insecticides, which have no residual activity.

Furthermore, one can distinguish three types of insecticide. 1. Natural insecticides, such as nicotine, pyrethrum and neem extracts, made by plants as defenses against insects. 2. Inorganic insecticides, which are metals. 3. Organic insecticides, which are organic chemical compounds, mostly working by contact.

The mode of action describes how the pesticide kills or inactivates a pest. It provides another way of classifying insecticides. Mode of action is important in understanding whether an insecticide will be toxic to unrelated species, such as fish, birds and mammals.

Insecticides may be repellent or non-repellent. Social insects such as ants cannot detect non-repellents and readily crawl through them. As they return to the nest they take insecticide with them and transfer it to their nestmates. Over time, this eliminates all of the ants including the queen. This is slower than some other methods, but usually completely eradicates the ant colony.

Insecticides are distinct from non-insecticidal repellents, which repel but do not kill.

Type of activity

Systemic insecticides become incorporated and distributed systemically throughout the whole plant. When insects feed on the plant, they ingest the insecticide. Systemic insecticides produced by transgenic plants are called plant-incorporated protectants (PIPs). For instance, a gene that codes for a specific Bacillus thuringiensis biocidal protein was introduced into corn and other species. The plant manufactures the protein, which kills the insect when consumed.

Contact insecticides are toxic to insects upon direct contact. These can be inorganic insecticides, which are metals and include arsenates, copper and fluorine compounds, which are less commonly used, and the commonly used sulfur. Contact insecticides can be organic insecticides, i.e. organic chemical compounds, synthetically produced, and comprising the largest numbers of pesticides used today. Or they can be natural compounds like pyrethrum, neem oil etc. Contact insecticides usually have no residual activity.

Efficacy can be related to the quality of pesticide application, with small droplets, such as aerosols often improving performance

Bacillus thuringiensis (or Bt) is a Gram-positive, soil-dwelling bacterium, commonly used as a biological pesticide. B. thuringiensis also occurs naturally in the gut of caterpillars of various types of moths and butterflies, as well on leaf surfaces, aquatic environments, animal feces, insect-rich environments, and flour mills and grain-storage facilities.

During sporulation, many Bt strains produce crystal proteins (proteinaceous inclusions), called ä-endotoxins, that have insecticidal action. This has led to their use as insecticides, and more recently to genetically modified crops using Bt genes, such as Bt corn. Many crystal-producing Bt strains, though, do not have insecticidal properties.

TAXONOMY AND DISCOVERY

B. thuringiensis was first discovered in 1901 by Japanese biologist Ishiwata Shigetane (ów!n A~ä€) in silkworms. He named it Bacillus sotto,using the Japanese word sottô (RSP, 'collapse'), here referring to bacillary paralysis. In 1911, German microbiologist Ernst Berlinerindependently rediscovered it when he isolated it as the cause of a disease called Schlaffsucht in flour moth caterpillars in Thuringia (hence the specific name thuringiensis, "Thuringian"). B. sotto would later be reassigned as B. thuringiensis var. sotto.

In 1976, Robert A. Zakharyan reported the presence of a plasmid in a strain of B. thuringiensis and suggested the plasmid's involvement in endospore and crystal formation. B. thuringiensis is closely related to B. cereus, a soil bacterium, and B. anthracis, the cause of anthrax; the three organisms differ mainly in their plasmids. Like other members of the genus, all three are aerobes capable of producing endospores.

Subspecies

There are several dozen recognized subspecies of bacillus thuringiensis. Subspecies commonly used as insecticides include Bacillus thuringiensis subspecies kurstaki (Btk) and subspecies israelensis (Bti) and subspecies aizawa (Bta).

Mechanism of insecticidal action

Upon sporulation, B. thuringiensis forms crystals of proteinaceous insecticidal ä-endotoxins (called crystal proteins or Cry proteins), which are encoded by cry genes. In most strains of B. thuringiensis, the cry genes are located on a plasmid (cry is not a chromosomal gene in most strains).

Cry toxins have specific activities against insect species of the orders Lepidoptera (moths and butterflies), Diptera (flies and mosquitoes), Coleoptera (beetles), Hymenoptera (wasps, bees, ants and sawflies) and nematodes. Thus, B. thuringiensis serves as an important reservoir of Cry toxins for production of biological insecticides and insect-resistant genetically modified crops. When insects ingest toxin crystals, their alkaline digestive tracts denature the insoluble crystals, making them soluble and thus amenable to being cut with proteases found in the insect gut, which liberate the toxin from the crystal. The Cry toxin is then inserted into the insect gut cell membrane, paralyzing the digestive tract and forming a pore. The insect stops eating and starves to death; live Bt bacteria may also colonize the insect which can contribute to death. The midgut bacteria of susceptible larvae may be required for B. thuringiensis insecticidal activity.

In 1996 another class of insecticidal proteins in Bt was discovered; the vegetative insecticidal proteins (Vip). Vip proteins do not share sequence homology with Cry proteins, in general do not compete for the same receptors, and some kill different insects than do Cry proteins.

In 2000, a novel functional group of Cry protein, designated parasporin, was discovered from noninsecticidal B. thuringiensis isolates. The proteins of parasporin group are defined as B. thuringiensis and related bacterial parasporal proteins that are not hemolytic, but capable of preferentially killing cancer cells. As of January 2013, parasporins comprise six subfamilies: PS1 to PS6.

Use of spores and proteins in pest control

Spores and crystalline insecticidal proteins produced by B. thuringiensis have been used to control insect pests since the 1920s and are often applied as liquid sprays. They are now used as specific insecticides under trade names such as DiPel and Thuricide. Because of their specificity, these pesticides are regarded as environmentally friendly, with little or no effect on humans, wildlife, pollinators, and most other beneficial insects, and are used in organic farming; however, the manuals for these products do contain many environmental and human health warnings, and a 2012 European regulatory peer review of five approved strains found, while data exist to support some claims of low toxicity to humans and the environment, the data are insufficient to justify many of these claims.

New strains of Bt are developed and introduced over time as insects develop resistance to Bt, or the desire occurs to force mutations to modify organism characteristics or to use homologous recombinant genetic engineering to improve crystal size and increase pesticidal activity or broaden the host range of Bt and obtain more effective formulations. Each new strain is given a unique number and registered with the U.S. EPA and allowances may be given for genetic modification depending on "its parental strains, the proposed pesticide use pattern, and the manner and extent to which the organism has been genetically modified". Formulations of Bt that are approved for organic farming in the US are listed at the website of the Organic Materials Review Institute (OMRI) and several university extension websites offer advice on how to use Bt spore or protein preparations in organic farming.

The Belgian company Plant Genetic Systems (now part of Bayer CropScience) was the first company (in 1985) to develop genetically modified crops (tobacco) with insect tolerance by expressing cry genes from B. thuringiensis; the resulting crops contain delta endotoxin. The Bt tobacco was never commercialized; tobacco plants are used to test genetic modifications since they are easy to manipulate genetically and are not part of the food supply.

Usage

In 1995, potato plants producing CRY 3A Bt toxin were approved safe by the Environmental Protection Agency, making it the first human-modified pesticide-producing crop to be approved in the USA,though many plants produce pesticides naturally, including tobacco, coffee plants, cocoa, and black walnut. This was the 'New Leaf' potato, and it was removed from the market in 2001 due to lack of interest. For current crops and their acreage under cultivation, see genetically modified crops.

In 1996, genetically modified maize producing Bt Cry protein was approved, which killed the European corn borer and related species; subsequent Bt genes were introduced that killed corn rootworm larvae.

The Bt genes engineered into crops and approved for release include, singly and stacked: Cry1A.105, CryIAb, CryIF, Cry2Ab, Cry3Bb1, Cry34Ab1, Cry35Ab1, mCry3A, and VIP, and the engineered crops include corn and cotton.

Corn genetically modified to produce VIP was first approved in the US in 2010.

In India, by 2014, more than seven million cotton farmers, occupying twenty-six million acres, had adopted Bt cotton.

Monsanto developed a soybean expressing Cry1Ac and the glyphosate-resistance gene for the Brazilian market, which completed the Brazilian regulatory process in 2010.

SAFETY STUDIES

The use of Bt toxins as plant-incorporated protectants prompted the need for extensive evaluation of their safety for use in foods and potential unintended impacts on the environment.

Dietary risk assessment

Concerns over the safety of consumption of genetically-modified plant materials that contain Cry proteins have been addressed in extensive dietary risk assessment studies. While the target pests are exposed to the toxins primarily through leaf and stalk material, Cry proteins are also expressed in other parts of the plant, including trace amounts in maize kernels which are ultimately consumed by both humans and animals.

Toxicology studies

Animal models have been used to assess human health risk from consumption of products containing Cry proteins. The United States Environmental Protection Agency recognizes mouse acute oral feeding studies where doses as high as 5,000 mg/kg body weight resulted in no observed adverse effects. Research on other known toxic proteins suggests that toxicity occurs at much lower doses, further suggesting that Bt toxins are not toxic to mammals. The results of toxicology studies are further strengthened by the lack of observed toxicity from decades of use of B. thuringiensis and its crystalline proteins as an insecticidal spray.

Allergenicity studies

Introduction of a new protein raised concerns regarding the potential for allergic responses in sensitive individuals. Bioinformatic analysis of known allergens has indicated there is no concern of allergic reactions as a result of consumption of Bt toxins. Additionally, skin prick testingusing purified Bt protein resulted in no detectable production of toxin-specific IgE antibodies, even in atopic patients.

DIGESTIBILITY STUDIES

Studies have been conducted to evaluate the fate of Bt toxins that are ingested in foods. Bt toxin proteins have been shown to digest within minutes of exposure to simulated gastric fluids. The instability of the proteins in digestive fluids is an additional indication that Cry proteins are unlikely to be allergenic, since most known food allergens resist degradation and are ultimately absorbed in the small intestine.

Ecological risk assessment

Ecological risk assessment aims to ensure there is no unintended impact on non-target organisms and no contamination of natural resources as a result of the use of a new substance, such as the use of Bt in genetically-modified crops. The impact of Bt toxins on the environments where transgenic plants are grown has been evaluated to ensure no adverse effects outside of targeted crop pests.

Persistence in environment

Concerns over possible environmental impact from accumulation of Bt toxins from plant tissues, pollen dispersal, and direct secretion from roots have been investigated. Bt toxins may persist in soil for over 200 days, with half-lives between 1.6 and 22 days. Much of the toxin is initially degraded rapidly by microorganisms in the environment, while some is adsorbed by organic matter and persists longer. Some studies, in contrast, claim that the toxins do not persist in the soil. Bt toxins are less likely to accumulate in bodies of water, but pollen shed or soil runoff may deposit them in an aquatic ecosystem. Fish species are not susceptible to Bt toxins if exposed.

Impact on non-target organisms

The toxic nature of Bt proteins has an adverse impact on many major crop pests, but ecological risk assessments have been conducted to ensure safety of beneficial non-target organisms that may come into contact with the toxins. Widespread concerns over toxicity in non-target lepidopterans, such as the monarch butterfly, have been disproved through proper exposure characterization, where it was determined that non-target organisms are not exposed to high enough amounts of the Bt toxins to have an adverse effect on the population. Soil-dwelling organisms, potentially exposed to Bt toxins through root exudates, are not impacted by the growth of Bt crops.

INSECT RESISTANCE

In November 2009, Monsanto scientists found the pink bollworm had become resistant to the first-generation Bt cotton in parts of Gujarat, India - that generation expresses one Bt gene, Cry1Ac. This was the first instance of Bt resistance confirmed by Monsanto anywhere in the world. Monsanto immediately responded by introducing a second-generation cotton with multiple Bt proteins, which was rapidly adopted. Bollworm resistance to first-generation Bt cotton was also identified in Australia, China, Spain, and the United States.

Secondary pests

Several studies have documented surges in "sucking pests" (which are not affected by Bt toxins) within a few years of adoption of Bt cotton. In China, the main problem has been with mirids, which have in some cases "completely eroded all benefits from Bt cotton cultivation".The increase in sucking pests depended on local temperature and rainfall conditions and increased in half the villages studied. The increase in insecticide use for the control of these secondary insects was far smaller than the reduction in total insecticide use due to Bt cotton adoption.Another study in five provinces in China found the reduction in pesticide use in Bt cotton cultivars is significantly lower than that reported in research elsewhere, consistent with the hypothesis suggested by recent studies that more pesticide sprayings are needed over time to control emerging secondary pests, such as aphids, spider mites, and lygus bugs.

Similar problems have been reported in India, with both mealy bugs and aphids although a survey of small Indian farms between 2002 and 2008 concluded Bt cotton adoption has led to higher yields and lower pesticide use, decreasing over time.

Controversies

There are controversies around GMOs on several levels, including whether making them is ethical, whether food produced with them is safe, whether such food should be labeled and if so how, whether agricultural biotech is needed to address world hunger now or in the future, and more specifically to GM crops—intellectual property and market dynamics; environmental effects of GM crops; and GM crops' role in industrial agricultural more generally. There are also issues specific to Bt transgenic crops.

Lepidopteran toxicity

The most publicised problem associated with Bt crops is the claim that pollen from Bt maize could kill the monarch butterfly. The paper produced a public uproar and demonstrations against Bt maize; however by 2001 several follow-up studies coordinated by the USDA had asserted that "the most common types of Bt maize pollen are not toxic to monarch larvae in concentrations the insects would encounter in the fields."

Wild maize genetic mixing

A study published in Nature in 2001 reported Bt-containing maize genes were found in maize in its center of origin, Oaxaca, Mexico. In 2002, paper concluded, "the evidence available is not sufficient to justify the publication of the original paper." A significant controversy happened over the paper and Nature's unprecedented notice.

A subsequent large-scale study, in 2005, failed to find any evidence of genetic mixing in Oaxaca. A 2007 study found the "transgenic proteins expressed in maize were found in two (0.96%) of 208 samples from farmers' fields, located in two (8%) of 25 sampled communities." Mexico imports a substantial amount of maize from the US, and due to formal and informal seed networks among rural farmers, many potential routes are available for transgenic maize to enter into food and feed webs. One study found small-scale (about 1%) introduction of transgenic sequences in sampled fields in Mexico; it did not find evidence for or against this introduced genetic material being inherited by the next generation of plants. That study was immediately criticized, with the reviewer writing, "Genetically, any given plant should be either nontransgenic or transgenic, therefore for leaf tissue of a single transgenic plant, a GMO level close to 100% is expected. In their study, the authors chose to classify leaf samples as transgenic despite GMO levels of about 0.1%. We contend that results such as these are incorrectly interpreted as positive and are more likely to be indicative of contamination in the laboratory."

Colony collapse disorder

As of 2007, a new phenomenon called colony collapse disorder (CCD) began affecting bee hives all over North America. Initial speculation on possible causes included new parasites, pesticide use, and the use of Bt transgenic crops. The Mid-Atlantic Apiculture Research and Extension Consortium found no evidence that pollen from Bt crops is adversely affecting bees. According to the USDA, "Genetically modified (GM) crops, most commonly Bt corn, have been offered up as the cause of CCD. But there is no correlation between where GM crops are planted and the pattern of CCD incidents. Also, GM crops have been widely planted since the late 1990s, but CCD did not appear until 2006. In addition, CCD has been reported in countries that do not allow GM crops to be planted, such as Switzerland. German researchers have noted in one study a possible correlation between exposure to Bt pollen and compromised immunity to Nosema." The actual cause of CCD was unknown in 2007, and scientists believe it may have multiple exacerbating causes.

OTHER BIOLOGICAL APPROACHES

Plant-incorporated protectants

Transgenic crops that act as insecticides began in 1996 with a genetically modified potato that produced the Cry protein, derived from the bacterium Bacillus thuringiensis, which is toxic to beetle larvae such as the Colorado potato beetle. The technique has been expanded to include the use of RNA interference RNAi that fatally silences crucial insect genes. RNAi likely evolved as a defense against viruses. Midgut cells in many larvae take up the molecules and help spread the signal. The technology can target only insects that have the silenced sequence, as was demonstrated when a particular RNAi affected only one of four fruit fly species. The technique is expected to replace many other insecticides, which are losing effectiveness due to the spread of pesticide resistance.

Enzymes

Many plants exude substances to repel insects. Premier examples are substances activated by the enzyme myrosinase. This enzyme converts glucosinolates to various compounds that are toxic to herbivorous insects. One product of this enzyme is allyl isothiocyanate, the pungent ingredient in horseradish sauces.

A genetically engineered potato is a potato that has had its genes modified, using genetic engineering. Goals of modification include introducing pest resistance, tweaking the amounts of certain chemicals produced by the plant, and to prevent browning or bruising of the tubers. Varieties modified to produce large amounts of starches may be approved for industrial use only, not for food. Innate

The genetically modified Innate potato was approved by the USDA in 2014 and the FDA in 2015. The cultivar was developed by J. R. Simplot Company. It is designed to resist blackspot bruising, browning and to contain less of the amino acid asparagine that turns into acrylamide during the frying of potatoes. Acrylamide is a probable human carcinogen, so reduced levels of it in fried potato foods is desirable. The 'Innate' name comes from the fact that this variety does not contain any genetic material from other species (the genes used are "innate" to potatoes) and uses RNA interference to switch off genes. Simplot hopes that not including genes from other species will assuage consumer fears about biotechnology.

The "Innate" potato is not a single cultivar, rather, it is a group of potato varieties that have had the same genetic alterations applied using the same process. Five different potato varieties have been transformed, creating "innate" versions of the varieties, with all of the original traits, plus the engineered ones. Ranger Russet, Russet Burbank, and Atlantic potatoes have all been transformed by Simplot, as well as two proprietary varieties. Modifications of each variety involved two transformations, one for each of the two new traits, thus there was a total of ten transformation events in developing the different Innate varieties.

McDonald's is a major consumer of potatoes in the US. The Food and Water Watch has petitioned the company to reject the newly marketed Innate potatoes. McDonald's has announced that they have ruled out using Innate.

PREVIOUSLY MARKETED VARIETIES

Used for food

NewLeaf

In 1995, Monsanto introduced the NewLeaf variety of potato which was their first genetically modified crop. It was designed to resist attack from the Colorado potato beetle due to the insertion of Bt toxin producing genes from the bacterium Bacillus thuringiensis. The insect-resistant potatoes found only a small market, and Monsanto discontinued the sale of seed in 2001.

Used in industry

Amflora

'Amflora' (also known as EH92-527-1) was a cultivar developed by BASF Plant Science for production of pure amylopectin starch for processing into waxy potato starch. It was approved for industrial applications in the European Union market on 2 March 2010 by the European Commission, but was withdrawn from the EU market in January 2012 due to a lack of acceptance from farmers and consumers.

Unmarketed varieties

A modified Désirée potato was developed in the 1990s by biochemist John Gatehouse at Cambridge Agricultural Genetics (later renamed Axis Genetics) and had gone through two years of field trials at Rothamsted Experimental Station. The potatoes were modified to express the Galanthus nivalis agglutinin (GNA) gene from the Galanthus (snowdrop) plant, which caused them to produce GNA lectin protein that is toxic to some insects. This variety of potatoes is the one which was involved in the Pusztai affair.

In 2014, a team of British scientists published a paper about three-year field trial showing that another genetically modified version of the Désirée cultivar can resist infection after exposure to late blight, one of the most serious diseases of potatoes. They developed this potato for blight resistance by inserting a gene (Rpi-vnt1.1), into the DNA of Désirée potatoes. This gene, which conferred the resistance to blight, was isolated from a wild relative of potatoes, Solanum venturii, which is a native of South America.

Bacillus thuringiensis is a bacterial disease that affects Lepidopterans and some other insects. Toxins produced by strains of this bacterium are used as a larvicide against caterpillars, beetles, and mosquitoes. Toxins from Saccharopolyspora spinosa are isolated from fermentations and sold as Spinosad. Because these toxins have little effect on other organisms, they are considered more environmentally friendly than synthetic pesticides. The toxin from B. thuringiensis (Bt toxin) has been incorporated directly into plants through the use of genetic engineering. Other biological insecticides include products based on entomopathogenic fungi (e.g., Beauveria bassiana, Metarhizium anisopliae), nematodes (e.g., Steinernema feltiae) and viruses (e.g., Cydia pomonella granulovirus).

SYNTHETIC INSECTICIDE

A major emphasis of organic chemistry is the development of chemical tools to enhance agricultural productivity. Insecticides represent a major area of emphasis. Many of the major insecticides are inspired by biological analogues. Many others are completely alien to nature.

Organochlorides

The best known organochloride, DDT, was created by Swiss scientist Paul Müller. For this discovery, he was awarded the 1948 Nobel Prize for Physiology or Medicine. DDT was introduced in 1944. It functions by opening sodium channels in the insect's nerve cells. The contemporaneous rise of the chemical industry facilitated large-scale production of DDT and related chlorinated hydrocarbons.

Organophosphates and carbamates

Organophosphates are another large class of contact insecticides. These also target the insect's nervous system. Organophosphates interfere with the enzymes acetylcholinesterase and other cholinesterases, disrupting nerve impulses and killing or disabling the insect. Organophosphate insecticides and chemical warfare nerve agents (such as sarin, tabun, soman, and VX) work in the same way. Organophosphates have a cumulative toxic effect to wildlife, so multiple exposures to the chemicals amplifies the toxicity. In the US, organophosphate use declined with the rise of substitutes.

Carbamate insecticides have similar mechanisms to organophosphates, but have a much shorter duration of action and are somewhat less toxic.

Pyrethroids

Pyrethroid pesticides mimic the insecticidal activity of the natural compound pyrethrum, the biopesticide found in pyrethrins. These compounds are nonpersistent sodium channel modulators and are less toxic than organophosphates and carbamates. Compounds in this group are often applied against household pests.

Neonicotinoids

Neonicotinoids are synthetic analogues of the natural insecticide nicotine (with much lower acute mammalian toxicity and greater field persistence). These chemicals are acetylcholine receptor agonists. They are broad-spectrum systemic insecticides, with rapid action (minutes-hours). They are applied as sprays, drenches, seed and soil treatments. Treated insects exhibit leg tremors, rapid wing motion, stylet withdrawal (aphids), disoriented movement, paralysis and death. Imidacloprid may be the most common. It has recently come under scrutiny for allegedly pernicious effects on honeybees and its potential to increase the susceptibility of rice to planthopper attacks.

Ryanoids

Ryanoids are synthetic analogues with the same mode of action as ryanodine, a naturally occurring insecticide extracted from Ryania speciosa(Flacourtiaceae). They bind to calcium channels in cardiac and skeletal muscle, blocking nerve transmission. Only one such insecticide is currently registered, Rynaxypyr, generic name chlorantraniliprole.

Insect growth regulators

Insect growth regulator (IGR) is a term coined to include insect hormone mimics and an earlier class of chemicals, the benzoylphenyl ureas, which inhibit chitin (exoskeleton) biosynthesis in insects Diflubenzuron is a member of the latter class, used primarily to control caterpillarsthat are pests. The most successful insecticides in this class are the juvenoids (juvenile hormone analogues). Of these, methoprene is most widely used. It has no observable acute toxicity in rats and is approved by World Health Organization (WHO) for use in drinking water cisterns to combat malaria. Most of its uses are to combat insects where the adult is the pest, including mosquitoes, several fly species, and fleas. Two very similar products, hydroprene and kinoprene, are used for controlling species such as cockroaches and white flies. Methoprene was registered with the EPA in 1975. Virtually no reports of resistance have been filed. A more recent type of IGR is the ecdysone agonist tebufenozide (MIMIC), which is used in forestry and other applications for control of caterpillars, which are far more sensitive to its hormonal effects than other insect orders.

ENVIRONMENTAL HARM

Effects on nontarget species

Some insecticides kill or harm other creatures in addition to those they are intended to kill. For example, birds may be poisoned when they eat food that was recently sprayed with insecticides or when they mistake an insecticide granule on the ground for food and eat it.

Sprayed insecticide may drift from the area to which it is applied and into wildlife areas, especially when it is sprayed aerially.

DDT

The development of DDT was motivated by desire to replace more dangerous or less effective alternatives. DDT was introduced to replace leadand arsenic-based compounds, which were in widespread use in the early 1940s.

DDT was brought to public attention by Rachel Carson's book Silent Spring. One side-effect of DDT is to reduce the thickness of shells on the eggs of predatory birds. The shells sometimes become too thin to be viable, reducing bird populations. This occurs with DDT and related compounds due to the process of bioaccumulation, wherein the chemical, due to its stability and fat solubility, accumulates in organisms' fatty tissues. Also, DDT may biomagnify, which causes progressively higher concentrations in the body fat of animals farther up the food chain. The near-worldwide ban on agricultural use of DDT and related chemicals has allowed some of these birds, such as the peregrine falcon, to recover in recent years. A number of organochlorine pesticides have been banned from most uses worldwide. Globally they are controlled via the Stockholm Convention on persistent organic pollutants. These include: aldrin, chlordane, DDT, dieldrin, endrin, heptachlor, mirex and toxaphene.

POLLINATOR DECLINE

Insecticides can kill bees and may be a cause of pollinator decline, the loss of bees that pollinate plants, and colony collapse disorder (CCD),in which worker bees from a beehive or Western honey bee colony abruptly disappear. Loss of pollinators means a reduction in crop yields.Sublethal doses of insecticides (i.e. imidacloprid and other neonicotinoids) affect bee foraging behavior. However, research into the causes of CCD was inconclusive as of June 2007

A Pesticide Pollution

From the term pesticides are a molecules chemical substance or a mixture of molecules chemical substances or other agents that control or destroy any organisms that are considered as a pest. Pesticides are used to increase the protection of food and fibre and to promote public health. There are many types and producer of pesticide, but the pesticide that kills insects is called insecticide and one that kill plants like weeds called an herbicide may other life forms pesticide called fumigants and the ones that kill fungi grow on plants and some ties animals called fungicides.

Pest infestations have been problem to humans for as many thousands of years as human have practiced agriculture. For long period pests, flies, rats, lice and many other insects threatened human health. For thousands of years people looked for means to rid their crops of the insects eating them, the weeds chocking them or the fungi are making them uneatable. Therefore people began to use sulphur, a chemical product still in use by organic gardeners, as a pesticide thousands of years ago.

Extracts of chrysanthemum flowers containing pyrethrum have been used for nearly as long, and tobacco extract containing nicotine have been used for hundreds of years. Starting in the 1800s, chemical pesticides containing arsenic, mercury, lead, and copper came in to widespread use. An elderly man wrote a letter to a periodical in 1989 describing his grandmothers 1920s gardening chemical; in addition to her occasional use of the highly toxic gas hydrogen cyanide as a fumigant she use Paris green (copper arsenate), lead arsenate and nicotine sulphate to control garden pest.

In the first half of this century has given the widespread use of metal pesticide.

The first household hazardous waste roundup that Massachusetts carried out in the 1980s, recovered tree tons of arsenic in chemical that had been sitting in sheds and barns for many years. In large amounts, sulphate and cooper only partially controlled pests. Therefore it is not surprising when the very effective synthetic insecticide dichlorodiphenyltrichloroethane (DDT) was introduced in 1942 it was quickly embraced. DDT was mortal to many insects. It killed the mosquitoes and flies that spread disease, the insect infesting crops, and other insect such as body lice.

It was considered a tremendous contribution to public health, and the discoverer of its insecticidal activity received a Noble Prize in medicine in 1948. Many other synthetic chemical pesticides were quickly developed and saw widespread use. In 1940s, the ability of insects to mutate and become resistant to pesticide was observed; however, most pesticides remained widly effective and the phenomenon of resistance cause little concern.

DDT and similar organochlorine pesticides showed relatively low acute toxicity to human and were not absorbed through the skin. Possible chronic toxicity was little considered. The result was wide and often indiscriminate use of pesticide. It was not until the early 1960s that Rachel Carson's famous book Silent spring forced Americans to see the darker face of DDT and other pesticides.

Since the dark face of pesticides discovered it become to be as death sentence of the world and started to be regulated to minim of use and more carful of the way to be use by label it and advertising of the instructed of use by many of the world governments up to our days.

Pesticide use in EU and UK

The European commission (EC) together with a proposal for a framework directive adopted in 2006 the thematic strategy on the sustainable use of pesticides, and they aim to fill all the gaps for the current legislative regarding the level of use for pesticides in EU by setting up some of minimums rules for the use of pesticide in the community to reduce the risks of the pesticides affect on the human health and the environmental.

Therefore the EC directive laws which are:

SET RESIDUE LIMITS FOR CERTAIN PESTICIDES.

Prohibit the placing on the market of certain plant protective products.

In UK most people's use pesticides for different purposes, but UK still have control over what comes on to the market , the MRLs and them monitoring. Common sense suggests that the best way to prevent problems is to stop them at source. That means preventing the wrong kind of products getting to market and being used. Clearly, this admirable public health principle has gone a little awry over pesticides. But the European commission is aware of this and Brussels is increasingly the location of some fairly bitter arguments about pesticides product.

Pesticide and their application

There are many applications for pesticide usages, Pesticides in Agriculture. The use of pesticides enhance the crops to grow at time and in places where could not otherwise be grown. Fruits and vegetables are on market year round not only because they can't be transported long distance from warmer climates, but because pesticides makes it Let's see how pesticide use in Agriculture as the use of makes it possible to grow them over longer growing season and in a greater number of locations. For example without fungicides, certain crops could not grow in locals or in seasons when fungi grow prolifically. The health advantages of fresh fruit and vegetable availability year round and their lowered cost make up for any human health risk posed by pesticides. Another public health benefit is reducing growth of fungi on treated crops, fungi which can produce very toxic chemicals. Pesticides make monoculture possible, which is a large tract of land, can be devoted to only one crop, for example, wheat, cotton, soybeans, or corn, season after season at one location. Without chemicals, the pests that attack a monoculture crop would build up until the crop could no longer be grown at that location. Pesticides also make it possible to store food product for long periods. After harvest, grain is fumigated to kill the insects and diseases causing organism infesting it. These organisms could otherwise multiply during the storage, destroying part or all of the grain. For similar reasons crops are fumigated before being transported long distance to market. Pesticides are also used to control the vector that spread disease, such as mosquitoes, flies, ticks, and rats.

Disinfectants germicides are uses to kill microorganisms' that live outside the body. Regulated as pesticides by EPA, disinfectants have been used since 1867, when Lister began using phenol to disinfect operating rooms.

Chemical related to phenol are still widely in use for disinfectants. The active infect ants use in home and industries are chlorine containing compounds such as sodium hypochlorite, household bleach. For the antibiotics that benne used to kill microorganisms in human and animals are regulated by the US FDA, not EPA 20

Mode of action

It is very important to understand how pesticides mechanism to deal and study the harmful side effects of them and very necessary to pests targeted system function. It is also helpful to understand how animal and humans systems roles or functions to see the similarities and differences between humans and pests to have better control. Another reason it is important to understand the mode of actions of the pesticides we use is to avert the development of pesticide ability and the aim that pests try to achieve. The pesticides with the same mode of achievement action provide to this problem by killing the easily affected pests and leaving only those with conflict to the entire category of pesticides that work through identical mechanisms. Growth of pest conflict can be avoid or deferred by turning pest chemical rule that effort throughout dissimilar mode of achievement Insecticides and miticides in general target the nervous system, expansion and improvement, or energy production of the pest. Pesticides can also cause danger to workforce for the period of production, transportation, or at some stage in and after use. Bystander may also be affected at time, for example walker using public and civil rights of way on nearby land or families whose homes are close by harvest spraying actions. One of the most important hazards of pesticide use is to farm workers and gardeners. A recent advanced study by the Harvard School of Public Health discovered a 70% raise in the risk of developing Parkinson's infection for people expose to constant small level of pesticides.

Organchlorine pesticide DDT acts on nerve membranes to prevent normal conduction of nervous impulse. Organophosphate insecticides inhabit the action of the enzymes that breaks down the neurotransmitter acetylcholine; as acetylcholine leaves build up, the results are uncontrolled firing of nerves. Campmate insecticides exert toxicity in a similar manner, but the toxic effect are shorter lived. There are many other ways that insecticides can kill target pests: for example the botanical insecticides (pesticides derived from plant) rotenone is a stomach and contact poison.

Herbicides fall in too many chemical groups. Some interfere with the normal function of plant cell membranes, other act on plant metabolism to cause abnormal growth, and still others inhabit the action of enzymes necessary to plant life.

Some pesticides are selective. They act against a limited group of organisms because they affect some aspect of metabolism specific to a limited number of plant animal, or microbes. Any chemical can be toxic in high enough doses. However, an herbicide that interacts with an enzyme found only in planets is less likely to harm birds, other animals, and humans. Other pesticides are broad-spectrum, affecting a wider range of organisms and more likely to pose a danger to no target species. Fumigants are an example; the fumigants hydrogen cyanide and methyl bromide affect biochemical respiration in many species. A fumigant is often deliberately used to kill a variety of pests, those infesting the grain stored in an elevator or a greenhouse, for example. Fumigants are also used to sterilize soil or seeds. They are often gases that can penetrate an enclosed space to do the job required of them.

THE CAUSES OF PESTICIDE POLLUTION

Pesticides are a source of pollution, affect land and water everywhere in particular. The trouble is massive and increasing. According to the USA geological survey in 1990, pesticide pollution has been found in most or every lakes, stream, lakes, municipals, and agricultures lands. Many other nations are affected badly as well in the world. As the rain off water wash the chemical products close by water source and most of the chemical products are pesticides as it have been use to enhance the growing in the farming fields and from horticultural land and house gardens, but the main source of exposure to pesticides for mainly people is all the way through diet

In these days most new pesticide are with awareness regulated by government commandment in major countries in the world. As we could see that in Europe and UK, and the EPA in USA conducts studies and licence for pesticides to be used. However all this regulated laws to minimise the use of pesticides can not control what happen because when the grower open a particular product and spray it over his farm without reading the label on the product and follow the guidelines. Therefore accident will happens and can not be controlled.

For the so toxic pesticides as most of the pesticides are toxic they are restricted to a licences and trained application, especially in USA it is infringement to apply any pesticides in any way that is not in agreement with label for that pesticides and it is a offence could end up the farmer in the jail and judge him that he been used the product intentionally. The toxic pesticides are off the record according to their toxicity in the majority countries in the world. Most sensitive pesticide poisonings result from disregarding the label route. So the most important advice for the ones that must use toxic pesticides is to read the label carefully and follow the instruction to the letter, and for anyone who is concerned about the toxic effect of toxic in the pesticides that been used in agriculture is to try and eat the organic foods and vegetable as they grow without toxic pesticides. And there are many strategies available to organic gardens and farms to be avoid attacks by pest.

The effects of toxic pesticides on our foods and vegetables and any other effects on lands and our health for us and our children which is for sure become more and more crucial

Behaviour and fate of pesticides in the environmental

All types of pesticide made to be released in to the environment, and when we release a pesticide in to the environment many things happen to the pesticides. For example herbicides and insecticides are applied over large area of agriculture fields and forest, and farmers nay apply them a dozen times or more during the growing season, less than a half of the pesticides actually reach the insect, weed or other pests. Most become a pollutant. Sometimes a foggy or rainy weather prevents pesticides from being airborne away from the point of application, posing a problem to those exposed to the trapped pesticides. Most pesticides are applying to the crops by spraying then they drift by air from the point of application such as lands. The largest amount of the sprayed pesticides settle on to land and water close to the point of application, but the smaller amount swept higher in to the atmosphere with the winds, can be carried thousands of miles . Certain polychlorinated pesticides detected in wilderness lakes. In the United state and Canada are not used certain polychlorinated pesticides but still they have in the country lakes and they assume that have been blown from some other countries such as Mexico or Latin American. Once soil and water become contaminated with these present pesticides, they may remain so for many years, especially in the northern locations, where cold weather and lake of intense sunlight prevent them from degrading. However agriculture lands are major nonpoint sources of pesticides, fertilizers, eroded soil, and manure. Runoff from lands to which pesticides have been applied is responsible for most surface water contamination with pesticides. Occasionally what happen is advantageous. Such as, the escape of some herbicides into the agriculture roots ground region can give you improved weed control, but most the time releasing pesticides into the environment are risky and harmful, as not the whole applied chemical reach the target place, overflows can shift an herbicide away from objective weeds. The chemical is wasted, weed control is reduced, and there is more chance of damage other plants and polluting soil and water. Or some of the pesticide may drift downwind and outside of the future application site. Many procedure affect what happen to pesticides in the environment. These processes include breakdown, transfer, adsorption, and degradation. Transfer includes process that moves the pesticide away from the target place. These include leaching, volatilization, runoff, spray drift, chemical breakdown, absorption and removal of crops. In the below diagram we could see all the procedures when the pesticides release in to the environment.

THE FATE OF PESTICIDE IN SOIL

Soil qualities affected by pesticides, because they reduce the biodiversity in the soil and kill the entire future pest with many other small organisms that do live in soil. Due to the pesticides action in soil the life will be killed off and the soil quality become poor. This has a knock on effect upon the retention of water, for the farmer particularly in the time of drought this become a serious problem and issues. In such time organic farmer found out to have yielded approximately 25-40% higher than conventional farm. Soil fertility could be affected in other ways too. When most active organisms killed off in soil, the complex interactions which result in good fertility crack down. As well known that our plants depend on millions of bacteria and fungi's to pass nutrients to their rootlets, and when these circulations are disrupts plants turn out to be more dependent upon correct dose of chemical fertilisers at usual interval. The fantastically rich interactions in healthy soil can not be fully replicated. Therefore our nutrition and the soil are comprised, and will get large shape of vegetables and fruits, but very watery, which often lake taste and for sure they will contain pesticides residues.another most important factors influencing the action and biodegradability of pesticides in the soil is their attraction for adsorption by soil organic substance. Soil organic substance made of decomposed plant litter dead animals remains and roots and excreta, and they variable in both chemical and physical composition. It is also possesses a selection of a chemical functional groups like carboxyl, hydroxyl, pheonolic and amines which could interact with pesticides.

Many pesticides molecules are non ionic and non polar and in general hydrophobic, organic substance provide significant site for their adsorption. Adsorption is the process a chemical movements from a liquid state to the solid state. Adsorptions of a pesticide onto organic matter make affect its behavior and destiny in the soil in a number of ways. It may make the pesticide physiologically motionless and more vulnerable to degradation by microbial achievement, and therefore less moveable in the soil and less level to defeat process such as leaching. In other situation, adsorption may improve mobility of the pesticide. In the soil clarification, dissolve organic substance or colloidal particulate substance can form complexes with most hydrophobic compounds, including pesticides, greatly increasing their mobility through the soil profile and therefore their vulnerability to leaching defeat. As on the soil surface, pesticides linked with organic substance are susceptible to soil erosion and the movement to water course as balanced load. This sediment may then be deposited and build up in streams and lakes where it may prove unsettling to the aquatic ecosystem and linked food chains. Soil erosion is in charge for the disappearance of many disqualified pesticides example aldrin and dieldrin in surface watercourse in the UK. The clay content of soils may also significantly control the fate of agricultural pesticides. Clay particles size less than two micrometer in particular alumina silicates minerals have two significant properties which explain their primary consequence in soil chemistry. They may have very large specific surface areas and hold a permanent negative electrical charge. This means they are of considerable importance in the adsorption of ionic and ionizable pesticides. Many of the triazine herbicides, for example, are weak bases in acid media and one of the amino groups may become protonated, therefore enhancing its adsorption by clays at low soil pH.

Pesticides in surface and ground waters

The potential of water to spread massive epidemics is a matter of public record. In the beginning of the 20th century typhoid fever and other enteric disease were major causes of death. Since about 1920, however, these enteric diseases have been contributed little to sickness and death in many developing countries. This remarkable record is a credit to water resource engineering. Water borne disease out breaks still occur from time to time but are usually the result of accidents commonly involving small or private supplies. Concern over water borne viral diseases is a result of increased water reuse by man intensifying the need to know more about enteric viruses.

Specific needs of knowledge exist about trace amounts of some potentially toxic chemical or excessive amounts of some common minerals in drinking water. Many of the possible contaminants are organic compounds. These come from chemicals used as automotive fuel additives, insecticides, detergents, lubricants, and from many other types of industrial production. Toxicological effects of water borne organics have been observed principally in connection with the chlorinated hydrocarbons and organic phosphorus compounds use has pesticides these substances may enter the water from runoff, irrigation return flow, air drifting, and by direct application for the control of algae.

Surface and ground water tend to persist pesticides for long period. Therefore the hazard from pesticides in water results both from direct effects, and from indirect effect because they may be concentrated biologically in mans food chain. Generally fish are more sensitive to pesticides and many serve as rough method for determining when chlorinated hydrocarbon pesticides contend of water is approaching a danger level. This needs to be established as a fact and care must be exercised to select fish of the proper sensitivity.

The identification and qualification of pesticides compounds in water which have possible effects on human health pose a critical problem. Therefore under the Groundwater Regulations, you must have authorisation from the Environment Agency to dispose of pesticide washings on your land.

PRACTICAL SOLUTIONS FOR PESTICIDE POLLUTION

Pesticides pollution is everywhere, in everything better living through pesticides has turned out to have a serious pollution downside.

Pesticides pollution has become a big problem in many countries around the world. Although there are very strong laws been setup as pesticides pollution solution to prevent further pesticides pollution from taking place, but there is a lot of works is still to be done. The records from the environmental protection agency those around 41% of rivers, lakes, streams are not safe to fish to swim in due to the pesticides pollution in water and many other water pollution sources. Many laws in place that offer pesticides pollution but they are not always effectively in forces, and very simple way to have pesticides pollutions would be to enforce the rules that have been already set up.

Additional pesticides solutions involve reducing the amount of manure and encouraging smarter agricultural practises by using a biodynamic. Also we could be more advertising to reduce the households pesticides and fertilizers to the minimums need of usage or could be stopping their use altogether.

Individuals can do a lot to help prevent pesticides pollution at becomes a death sentence in the world and to aid in the pesticides pollution solution. Also we have to start buy and use organic food and green house hold cleaners and personal care items to prevent the run off of the chemical product into the ground water.

6

HOST PLANT RESISTANCE

Host plant resistance (HPR) to insects is an effective, economical, and environment friendly method of pest control. The most attractive feature of HPR is that farmers virtually do not need any skill in application techniques, and there is no cash investment by the resource poor farmers. Considerable progress has been made in identification and development of crop cultivars with resistance to the major pests in different crops. There is a need to transfer resistance genes into high-yielding cultivars with adaptation to different agro-ecosystems. Resistance to insects should form one of the criteria to release varieties and hybrids for cultivation by the farmers. Genes from the wild relatives of crops, and novel genes, such as those from Bacillus thuringiensis can also be deployed in different crops to make HPR an effective weapon to minimize the losses due to insect pests. HPR will not only cause a major reduction in pesticide use and slowdown the rate of development of resistance to insecticides in insect populations, but also lead to increased activity of beneficial organisms and reduction in pesticide residues in food and food products. One IPM strategy is to grow crop varieties that are resistant to pest damage. Crop varieties that are resistant to pest damage are said to have host plant resistance. Host plant resistance can be broken down into three categories: non-preference, antibiosis, and tolerance.

NON-PREFERENCE (ANTI-XENOSIS)

Host plants that express non-preference affect the way an insect pest perceives the desirability of the host plant. Non-preference plants either provide stimuli that are unattractive to the pest (color, odor, texture such as downy hairs) or fail to provide stimuli that are attractive to the pest. In this way, non-preference plants affect the behavior of pests.

Antibiosis

Antibiosis is a type of resistance in which the host plant causes injury, death, reduced longevity, or reduced reproduction of the pest. Often both a resistant and susceptible variety will have the same basic response to a pest, but the resistant variety will respond more quickly or more dramatically than the susceptible variety, reducing the amount of damage the pest causes. Plants that express antibiosis affect the biology of pests.

Host Plant Resistance and Tolerance to Insect Pests

Crop varieties resistant to insects are far less common than disease-resistant varieties, because plant breeders have traditionally focused more on disease resistance. However, if they are available, resistant varieties can be an effective defense against insect pests. But even when insect-resistant cultivars are not available, some varieties may be less attractive to pest species or may tolerate more damage than others. Plant size, shape, coloration, leaf hair, cuticle thickness, and natural chemicals (attractants and repellents) can all affect pest susceptibility. Farmers can do their own breeding by collecting non-hybrid seed from healthy plants in the field. Plants well adapted to local conditions will be more likely to resist pests.

Over the years, plants have developed a variety of resistance mechanisms to deter feeding and damage by insects. Insects, at the same time, have evolved ways of circumventing the defense mechanisms in plants. This interrelationship is dynamic and ongoing. Some plants are naturally pest-resistant and scientists are researching these varieties in hopes of finding new methods of pest control to reduce pesticide use. Some plants defend themselves through morphological and physiological mechanisms. Plant breeders are using this information to develop pest-resistant plant cultivars.

The development of pest resistant cultivars is done through conventional breeding, where resistant traits are selected and incorporated into breeding lines over multiple generations, and also through genetic engineering where genes for pest resistance are artificially introduced into the plant genome. Keep in mind that genetically modified seeds and transplants are not allowed under the USDA National Organic Program Rule.

Plants with resistance based on morphological characteristics have plant structures or characteristics that interfere with insect movement, feeding, or reproduction on or in the plant. Examples include color, thickness of the cell walls and plant tissue, surface wax, spines and trichomes (hairs). These features can make the plant less attractive visually, or present formidable physical barriers to pests. Thick cell walls and plant tissue help a plant to resist the tearing action of chewing mouthparts and the penetration of an insect stylet or ovipositor.

Plant leaves are protected against desiccation, insect feeding and disease by a layer of surface waxes over the cuticle. Wax may physically inhibit the movement of an insect across a leaf surface. The wax on leaf surfaces of some raspberry varieties contains chemicals that negatively affect certain insects. Researchers have shown that survival of diamondback moth (DBM) larvae is lower on cabbage varieties with glossy or waxy leaves than on non-glossy varieties. They found that the wax layer inhibited leaf mining by DBM larvae which also made them more susceptible to predators.

Trichomes are hairlike epidermal outgrowths, sometimes referred to as pubescence, that occur on leaves, shoots, and roots. Their impact on insect pests depends on trichome density, erectness, length, and shape. Some insects experience difficulty feeding on and ingesting plants with these small hairs. Trichomes may also release sticky substances that can trap or inhibit movement of small insects.

A few examples of pest arthropods that are at least partially suppressed by plant pubescence include bean aphid and potato leaf hopper on beans, two spotted spider mite on strawberry, and whitefly on tomatoes, peppers, and potatoes.

Other plant defenses involve the production of plant chemicals that affect insect physiology. Scientists have long wondered why some plants are more resistant to insects than others. In answering this question, they have identified at least 30,000 plant chemicals that are involved with defense against insect attack. One common compound, known as pyrethrum, is derived from the blooms of the African Chrysanthemum, Tanacetum cinerariaefolium.

Ecological Results of Agriculture: In contrast to natural ecosystems, most agricultural crop production systems are ecologically unstable, non-sustainable, and energy dependent. Man has through plant domestication and cultivation practices interfered in many ways with species diversity and natural defense mechanisms of plants. Cultivated crops originated from genetically diverse plant types. However, crop plants are now grown in large, genetically homogeneous stands, a practice that decreases genetic and species diversity and increases the likelihood of economically significant insect pest infestations. Defense mechanisms of plants are re-created in resistant plants. Plant defense mechanisms include escape in space and time, incompatible biological associations, physically and chemically derived barriers, and accommodation by replacement or repair of damaged plant parts. By re-creating plant defenses, genetic resistance to insect pests plays, in an environmentally compatible manner, a vital role in the attempt to enhance ecological stability in agricultural crops.

An IPM Direct Control Tactic: Plant resistance to insects is one of several cultural control methods. Cultural control methods involve use of agronomic practices to reduce insect pest abundance and damage below that which would have occurred if the practice had not been used. In IPM, plant resistance to insects refers to the use of resistant crop varieties to suppress insect pest damage. Plant resistance is intended to be used in conjunction with other direct control tactics.

Definition of an Insect-Resistant Plant: Definitions of an insect-resistant plant are many and varied. In the broadest sense, plant resistance is defined as "the consequence of heritable plant qualities that result in a plant being relatively less damaged than a plant without the qualities." In practical agricultural terms, an insect-resistant crop cultivar is one that yields more than a susceptible cultivar when confronted with insect pest invasion. Resistance of plants is relative and is based on comparison with plants lacking the resistance characters, i.e., susceptible plants.

Effect of Insect Pest-Plant Host Relationship: Insect-resistant crop varieties suppress insect pest abundance or elevate the damage tolerance level of the plants. In other words, insect-resistant plants alter the relationship an insect pest has with its plant host. How the relationship between the insect and plant is affected depends on the kind of resistance, e.g. antibiosis, antixenosis (non-preference), or tolerance.

Antibiosis resistance affects the biology of the insect so pest abundance and subsequent damage is reduced compared to that which would have occurred if the insect was on a susceptible crop variety. Antibiosis resistance often results in increased mortality or reduced longevity and reproduction of the insect.

Antixenosis resistance affects the behavior of an insect pest and usually is expressed as non-preference of the insect for a resistant plant compared with a susceptible plant.

Tolerance is resistance in which a plant is able to withstand or recover from damage caused by insect pest abundance equal to that damaging a plant without resistance characters (susceptible). Tolerance is a plant response to an insect pest. Thus, tolerance resistance differs from antibiosis and antixenosis resistance in how it affects the insect-plant relationship. Antibiosis and antixenosis resistance cause an insect response when the insect attempts to use the resistant plant for food, oviposition, or shelter.

Advantages to the Use of Insect-Resistant Crop Varieties: Use of insect-resistant crop varieties is economically, ecologically, and environmentally advantageous. Economic benefits occur because crop yields are saved from loss to insect pests and money is saved by not applying insecticides that would have been applied to susceptible varieties. In most cases, seed of insect-resistant cultivars costs no more, or little more, than for susceptible cultivars. Ecological and environmental benefits arise from increases in species diversity in the agroecosystem, in part because of reduced use of insecticides. Increases in species diversity increase ecosystem stability which promotes a more sustainable system far less polluted and detrimental to natural resources.

The IPM concept stresses the need to use multiple tactics to maintain insect pest abundance and damage below levels of economic significance. Thus, a major advantage to the use of insect-resistant crop varieties as a component of IPM arises from the ecological compatibility and compatibility with other direct control tactics. Insect-resistant cultivars synergize the effects of natural, biological, and cultural insect pest-suppression tactics. The "built-in" protection of resistant plants from insect pests functions at a very basic level, disrupting the normal association of the insect pest with its host plant. The compatible, complementary role plant resistance to insect pests plays with other direct control tactics is, in theory and practice, in concert with the objectives of IPM. All crop cultivars should contain resistance to insect pests.

Plant resistance to insect pests has advantages over other direct control tactics. For example, plant resistance to insects is compatible with insecticide use, while biological control is not. Plant resistance to insects is not density dependent, whereas biological control is. Plant resistance is specific, only affecting the target pest. Often effects of use of insect-resistant cultivars are cumulative over time. Usually the effectiveness of resistant cultivars is long-lasting.

The role of plant resistance to insects in IPM has been well defined, at least in theory. However, the specific role a resistant cultivar plays in a particular IPM situation is crucial to successful deployment of the resistant cultivar. The impact of the resistant cultivar on standard cultural, biological, and insecticidal control methods should be well defined. Likewise, the impact of each of these control tactics on the resistant cultivar also must be defined.

Several definitions have been used to convey the relative level of resistance in a plant. However, the problem of quantifying resistance continues to be a problem influencing farmer acceptance of insect-resistant cultivars. A better way to define resistance levels in agronomically improved resistant cultivars is through quantified comparisons of insect pest damage or plant yield loss of susceptible cultivars. Once insect pest abundance or damage to yield-loss relationships have been determined, economic threshold levels can be determined and combined with factors such as crop value and insect pest control costs to develop dynamic thresholds for use by producers. Dynamic thresholds provide a description of resistance and can reduce crop loss risk because limitations are known and remedial action can be taken when necessary. By using this system to define relative differences in insect pest resistance between cultivars, it may be possible to simply indicate that a resistant cultivar has a higher economic threshold level than a traditional susceptible cultivar.

PARASITIC PLANTS AND HOST RELATIONSHIPS

Survival of one life form is dependent not only on its ability to search and utilize available life sustaining resources but also upon its healthy competition for various such requisites with other life forms belonging to either the same or different species. To facilitate this, different life forms have evolved with special assets of abilities to exist, work and exhibit a tendency to proliferate so as to flag their dominance on ecological niche. Scientific communities at times are amazed by the kinds of interaction in which different life forms co-exist with each other and also with the surroundings. In plants, interaction science has emerged as a major area of research which deals with investigating and studying such phenomena in detail, their mechanism, reasoning for interactions of plant with plant, animals, and/or micro-organism. Evolution of a new species from an existing ancestral mob assures adaptation to new and more efficient modes of existence, wherein lesser inputs are invested for more possible benefits. One such example is the parasitic plants or epiphytes.

Parasitic plants comprise 19 families in angiosperms and constitute to about 4,100 species (Nickrent. and Musselman, 2004). Among these with various parasitic habits, mistletoes are well known as perrenial, polyphyletic group of angiospermic aerial plant parasites (Devkota, 2005) infecting generally host stem and/or branches, and belong to families, Viscaceae and Loranthaceae.

Mistletoes are considered as an important component of biodiversity (Watson, 2001; Shaw et al., 2004), and have been appraised not only due to their unique parasitic relationship with their respective hosts, their growth habits, mechanism of seed dispersal but also because of their miraculous medicinal values which includes their use as sympathetic medicine to take down abnormal tissue growth under cancer (Kunwar et al.2005).

They are predominantly known to occur on fruit trees Mistletoes are characterised by the development of a root like absorptive organ called haustorium which forms the host parasite interface and draws nutrients from the host conductive tissues to the parasite (Devkota, 2005).

Dendrophthoe falcata is a hemi-parasitic plant belonging to the Loranthaceae mistletoes and is considered as the commonest of all other mistletoes widely occurring in India (Parker acnd Westwood, 2009). The genus Dendrophthoe comprises of about 31 species spread across tropical Africa, Asia, and Australia (Flora of China, 2003) among which 7 species are found in India. The sp. falcata has its hold in India since ancient times and has been in use as traditional folklore medication for various ailments. Though a parasite incurring huge losses to commercial fruit and timer production, it is considered and welcomed as a sacred sign of prosperity in few areas of rural India. Well globally, much of the advanced scientific investigations have centered on other mistletoes such as Orobanche, Viscum etc. These studies have aimed towards understanding the parasite behaviour, sap flow dynamics at the host parasite interface, chemical compositions of floral organs, designing successful control strategies, evaluating medicinal potentials etc.

The present review will focus on the existing literature earlier attempts aiming study of various scientific aspects associated with D. falcata. The review considerably outlines the variety of studies performed revolving around the host-parasite relationship, morphology and geographical distribution, and studies aiming towards in vitro establishment and attempts towards scientific justification of the ethno-medicinal properties, and various existing and possible control strategies for this mistletoe family member.

Morphology and biogeography:

Evolutionarily, mistletoes belonging to Loranthaceae are considered older than those of Viscaceae and ever since their origin from dense Gondwanan forests; they have successfully encompassed tropics, subtropics and arid portions on the globe through their nutritional competition-driven conquests (Devkota et al, 2005). Notes on floral morphology and embryology in D.falcata have been provided by Singh (1952).

D. falcata bears grey barks, thick coriaceous leaves variable in shape with stout flowers (Wealth of India. 2002). The flowering inflorescences in D. falcata was previously referred to as axillary or as developing on the scars of fallen leaves, but Y.P.S Pundir (1996) verified it to be of strictly cauliflorous nature and also notified that it shares fundamental similarity to that of Ficus glomerata, F. pomifera and F. hispida.

Two of its varieties are widespread in India namely, var. falcata (Honey Suckled Mistletoe) and var. Coccinea (Red Honey Suckled Mistletoe) distinguished by occurrence of white and red flowering, respectively (Flowers of India, a World Wide Web resource). A comprehensive list of almost all the species within the genus Dendrophthoe can be availed (from web source at http://www.naturekind.org/taxpage/0/binomial/Dendrophthoe_falcata).

Hosts and host-parasite relationships: Mistletoes occur mostly on forest, fruit and ornamental host trees preferentially harboring zones rich in biodiversity and thus are found excessively on mountain ridges provided with favorably optimum light intensities and in few in slopes and plains (Devkota, 2005 and references therein). Mistletoes do not follow a uniform pattern of distribution which is affected by local environments and effected by habits of seed dispersing avian visitors.

Host range: Earliest, B.Singh (1962) reported that the leafy mistletoe D. falcata is parasitic on over 300 hosts in India, Pakistan, Southeast Asia, and northern Australia. India¿½s rich plant biodiversity in 1977 could report exclusive host range numbering 345 plants susceptible to infection from this mistletoe (Siva Sarma and Vijayalakshmi 1977) which is a significant number as it contributes to its comparatively recent global record for 401 hosts (Shaw 1993). Till date, D. falcata represents the only known mistletoe with the largest global host range (Calvin and Wilson, 2009) which is continuously and rapidly widening.

Haustoria: Mistleoes are known to form haustorial structures at the point of attachment to the host.

Among angiosperms, parasitic relationship through the formation of haustorial linkages is known to be widespread (Wilson and Calvin, 2006). In general, haustorial connections among 72 (of the 75) ariel parasitic genera may belong to either of the four types viz., epicortical roots (ERs), clasping unions, wood roses, and bark strands (Calvin and Wilson, 1998). ERs may run along the host branches in either direction forming haustorial structures at variable intervals while ¿½unions¿½ occur as single points of attachment of individual parasites hence pronounced as solitary. In D, falcata on different hosts two of the haustorial kinds have been observed viz., solitary unions as on Sugar apple (Annona squamosa), and epicortical roots as on sugar apple (Achras sapota), guava (Psidium guajava), pomegranate (Punica granatum) have been observed. It is unknown about what factors decide formation of different haustorial types by the leafy mistletoe on different hosts.

The host branches infected with D. falcata show a gradual reduction in growth and diameter as compared to other healthy uninfected branches (Karunaichamy et al, 1999).

It has also been reported that the development of mistletoe plants on the host tree is a dynamic process which necessarily leads to the death of the host tree and that the whole process may last for about a decade (Karunaichamy et al, 1999). It is known that mistletoes have higher nutrient titer than their host (Lamont, 1983; Karunaichamy et al, 1993) and this could probably be attributed to a competitively higher water (including dissolved nutrient and growth promoting metabolite) uptake rate by the parasite at its haustorium at a point in the host branch in comparison to that in the protruding lateral ends of the branch (that extends after the haustorium). This might render key nutrients be unavailable to the host branches that could repress growth post-infection (Stewart and Press, 1990). But in work by Karunaichamy et al (1999) one D.falcata seed per stem of Azadirachta indica seedling has been tested which has shown the life threatening effect on whole plant which again itself is at an immature seedling stage (so death of the whole plant could be expected). In a broad sense, physiological and metabolic perturbations induced by the parasite in the whole host plant outlined at the seedling stage surmised by the reduced host leaf area, leaf number, growth performance upon infection cannot be correlated to that in a fully mature host.

Once more, a report by Sridhar and Rama Rao (1978) unraveled the infection of D. falcata surprisingly hosting itself directly on the fruit of Acharas sapota (Sapota). Almost all the hemi-parasitic members of the Loranthaceae tap the xylem vessels of their host to avail water and minerals but to a considerable extent produce their own supply of assimilates (Kuijt, 1969). D. falcata does not have an indigenous rooting system and is dependent on the host for water and minerals. Nutrient dynamics have shown that a higher titre of N, P , K, Mg and Na in the leaves of mistletoe than the leaves of uninfected and infected hosts which may be due to differential translocation of elements within the host phloem (Prakash et al, 1967; Karunaichamy et al, 1999). Nitrogen loss is well pronounced in mistletoe infected hosts and a higher potassium levels in mistletoe is an indicative of higher transpiration rates and a lack of re-translocation system (Karunaichamy et al, 1999 and references therein). Though outlined above that the total phosphorous in mistletoe leaves is present in excess as compared to that the host leaves, fractionation studies have elucidated that percentage acid-insoluble phosphate is comparatively lesser in the parasite and also that further fractionation again infers a lower percentage of phospholipid, RNA, DNA and phosphoproteins. The DNA content of the infected host leaves however demonstrate a dramatic increase compared to the healthy uninfected leaves (Prakash et al, 1967).

Mistletoes are believed to mimic their hosts in floral structures. At some instance it was supported by a hypothesis which explains that such behaviour imparts protection to the mistletoe from the herbivores. Later, the hypothesis was field tested, rejected and replaced by a belief that a host parasite resemblance might help to safeguard hosts¿½ individual fitness by signaling the birds to expel mistletoe seeds on other uninfected hosts (Atsatt, 1979). Moreover, the hosts and parasite follow individual developmental scheduling of events under which the parasite might experience an influence from a firm epigenetic hormonal control on leaf size and shape thus evolving the mimicry or resemblance to the host.

List of hosts: Loranthaceous parasites were known in India since 1885 and until then only 35 species were known. Specifically for D. falcata from among the earliest of the reports, 268 host species have been known to be under the clutch of the deadly hemiparasite (Ravindranath and Narsimha Rao, 1959), and shortly later B.Singh (1962) reported its parasitic relation with over 300 hosts reported in India, Pakistan, Southeast Asia, and northern Australia. Until 1993, a report by Shaw III depicts a global increase of host range to 401. In Indian tropics 37 species of tree have been reported as the favourite hosts for D.falcata (Ghosh et al., 2002). Due to the ever increasing host range (Sridhar and Rama Rao, 1978) currently it is difficult to draw a margin between the more and less common host genera.

VECTORS FOR SEED DISPERSAL AND POLLINATION

Seed dispersal as well as pollination is usually mediated by the birds that thrive on fruits from the parasite and/or host. Particularly in southern India, Tickel¿½s Flowerpecker which is also named as Pale-billed Flowerpecker (Dicaeum erythrohynchos, Latham) is reported to facilitate seed dispersal of D. falcata among Neem (Azadirachta indica A. Juss. belonging to Meliaceae) through fecal excretions or regurgitations (Karunaichamy et al, 1999; Hambali, 1977 and references therein). The mistletoe bird also eats insects and therefore has a grinding gizzard. When eating the mistletoe berry the bird is able to close the gizzard and the toxic seeds are usually swallowed as a whole and are shunt through their gut in about 3-4 minute (Murphy et al., 1993) and because the seed has a sticky coating, to get rid of the seed the bird applies its vent to the edges of the host tree branches and may turn around whereby the seed then sticks onto the branch where it may subsequently germinate (Ali, 1931; Ali 1932; Davidar, 1985). These birds also act as vectors for pollination in the hemi-parasite whose flowers bear a mechanism that causes pollen to explosively spray on the plumage of the visiting flowerpeckers (Karunaichamy et al., 1999; Vidal-Russell and Nickrent, 2008). Studies conducted in the higher altitudes of Western Ghats (where both the mistletoes and the flowerpeckers occur predominantly), which parallel the western coast of India infer that the flowerpecker pollinated mistletoes have particularly developed feature specialized to attract a unique vector both to facilitate pollination and seed dispersal: the fruit and flowers have similar resemblance and more significantly, the fruiting time overlap with the next flowering season (Davidar, 1983). Hair-crested Drongo or Spangled Drongo (Dicrurus hottentottus) is an Asian bird of the family Dicruridae and sunbirds (family=Nectariniidae) also known to feed on the nectar from the D.falcata flowers also adds to the list of pollinators to this mistletoe (Kunwar et al, 2005 and references therein).

Phytochemistry: Results for the preliminary tests in phytochemical screening analyses using ethanolic extract of the whole plant reports the presence of carbohydrates, glycosides, steroids, tannins & phenolic compounds, flavonoids and triterpenes (Sahu et al, 2010). Chromatography of D.falcata fruits has recently added new members to the list of naturally occurring terpenes which has taxonomically been considered significant (Mallavadhani et al., 2006). Total phenolics in mistletoe have a comparatively higher concentration but it failed to correlate to any resistance by the host or its susceptibility to infection by the hemiparaste (Khanna et al,, 1968). Later a phenolic inhibitor of the enzyme a-glucan phosphorylase was reported ((Khanna et al,, 1971).

A study by Ramchandra and coworkers (2005) indicated that an elevated antioxidant potential in the mistletoe is due to sequestration of phenolic compounds from the host plants. Their HPTLC data on samples from mistletoe hosted on Mangifera indica (Mango), reportedly contained mangiferin, a C-glucosyl xanthone synthesized by the host plant. The report raises query concerning further investigations for whether the transfer of phenolics to the hemi-parasite indicates a preliminary host mediated defense or a mode of host identification by the parasite or as a mechanism from within the parasite to bypass rejection from the host plant.

In mature seed embryo 0.65 mg chlorophyll/g fresh wt; chlorophyll a : b ratio of 1.6 per mg chlorophyll has been evidenced. Moreover, in context to titre of enzymes for photorespiration, phosphoglycolate activity in the embryo has been recorded to be 16.7 ¿½mol substrate/min and 3-phosphoglycerate phosphatase activity 6.8 ¿½mol substrate/min. Similarly, NADH-linked glyoxylate reductase activity has been demonstrated in the embryo (0.09 ¿½mol substrate/min/g fresh wt) by Kachru, and Krishnan (1979). Chroamatography of the ethanolic extract followed by NMR and Mass spectrometry could isolate and identify Quercitin, a rhamnoside (Shihab et al, 2006). Hexokinase has been partially isolated and studied for its properties from D.falata leaves (Baijal and Sanwal, 1976).

In a comparative evaluation of this parasitic plant grown on various host plants, Nair et al.10) reported the accumulation of kaempferol, quercetin, myrecitin, and their glycosides.

Medicinal Properties: D.falcata possesses remarkable potentials as a medicinal plant evident from the wound healing, anti-microbial, anti-oxidant, antinociceptive properties of its ethanolic extracts (Pattanayak and Sunita, 2008, Shihab et al, 2006). Medicinal properties of this hemiparasite may vary in effects respective to different hosts it establishes a relation with (Mallavadhani et al., 2006).

The whole plant is used in indigenous system of medicine as cooling, bitter, astringent, aphrodisiac, narcotic and diuretic (Alekutty e al., 1993) and is useful in treating pulmonary tuberculosis, asthma, menstrual disorders, swelling wounds, ulcers, renal and vesical calculi and vitiated conditions of kapha and pitta (Anarthe et al., 20081; Anarthe et al, 2010; Sastry, 1952; Pattanayak et al., 2008). Also, the decoction of plant used by women as an anti-fertility agent has been evidenced to possess anticancer activity (Nadkarni, 1993). The leaf ethanolic extract significantly and dose dependently inhibits the acetic acid induced writhing in mice (Shihab et al, 2006) and has indicated a low level toxicity in the brine shrimp lethality assays. Besides, a more recent work by Pattanayak and Mazumder (2010) shows significant tumor reduction in induced mammary carcinogenesis in Wistar female rats when fed with hydroalcoholic extracts of D. falcata.

Ethno-medicinal aspects: Ethnomedical studies are generally characterized by a strong anthropological and/or biomedical approach and in this context the use of D.falcata as a traditioal remedy for most if not all the ailments has been well appraised. Perhaps, the variety of knowledge available has not yet been exploited full fledgedly due to the currently over-recommended chemial analogs that are predominant over market most of which have side-effects associated under frequent use.

In vitro regeneration and gene introgression studies: It is known that long term caulogensis from the endosperm tissue exhibits a high degree of polyploidisation of its cells associated with abnormalities such as chromosome bridges and lagging chromosomes but mistletoe endosperms have exceptional stability in their chromosome number (Razdan, 2003).

In the past, in vitro studies aiming towards investigating behavior of mature embryos have been reported in D.falcata (Johri and Bajaj, 1962). The effects of light, endosperm, endosperm extract, and position of the embryo on media have been studied on in vitro germination of D.falcata (Bajaj, 1968). Shoot induction has been successfully achieved in D.falcata (Bhojwani and Razdan, 1983). Shoot buds in Loranthaceae directly differentiate from the peripheral cells of the endosperm cultures though D.falcata may particularly require media supplemented with a combination of cytokinins and a low concentration of IAA or IBA. Among other cytokinins, 2-ip has been most effective and also addition of Casein hydrolysate (2000mg/l) has been necessarily recommended for shoot bud differentiation. For optimal shoot bud growth frequent subcultutres on fresh media is crucial. Caulogenesis from these shoot buds may be followed by differentiation into shoot and haustoria. The latter could be avoided by increased kinetin concentration and for such kinetin induced shoot bud differentiation, a low concentration of auxin (IAA or IBA) is necessary (Nag and Johri, 1971).

Advanced molecular level studies are surprisingly lacking for D.falcata with only single report where tests for amenability of this plant in gene introgression studies have been investigated via successful agrobacterium mediated transformation of its embryogenic cultures for GUS expression (Rohini et al, 2005).

Diseases of D. falcata: Very little is known about diseases that victimize this hemiparasitc plant. The pathogen species that have been reported in association are generally known to commonly infect a wide range of plants off genera. There are realizations that propose that mistletoe specific pathogens are rare and develop endemism and be localized to favorably optimized areas where their growth requisites and nutritional demands are inline with the host mistletoe species and their host nutrient and sap flow dynamics.

Fungal

The hemi-parasite itself is susceptible to diseases such as leaf blight caused by Colletotrichum state of Glomerella cingulata which has been reported on D.falcata which seriously infects Teak (Tectona grandis) plantations (Ali and Florence, 1987) in Kerala, India.

Epiparasitism

Plant: A surprising evidence of hyper-parasitism has been reported in the Mohand Range Forest (Dehradun, India) where D.falcata itself receives haustoria from Scurrula cordifolia (another mistletoe), both hosted together by Ougeinia oojeinensis (fabaceae) (Pundir, 1979). Similaly, Viscum orientale (Viscaceae) has also been reported to grow on D.falcata (Saxena, 1971). In another instance Cuscuta reflexa (Convolvulaceae) has been shown to act as a rival to the leafy mistletoe (Nath and Indira, 1975). Viscum capitellatum. Sm has also been evidenced from Sri Lanka to be an obligate epiparasite for D.falcata (Calvin and Wilson, 2009) and symplastic continuity exists between their phloem (Glatzel and Balasubramaniam, 1987) which is an indicative of the aggressive epiparasitism depicted more vividly by the presence of graft-resembling ¿½unions¿½. Viscum loranthi Elmer, thus has been advocated to have potentials for biological control of Loranthaceae hemiparasites (Pundir, 1981) damaging commercial timber crops. Well, other species such as D.glabrata has also been reported to be parasitized by Viscum articulatum (Robert L. Mathiasen, David C. Shaw, Daniel L. Nickrent, David M. Watson,2008 Plant Disease / Vol. 92 No. 7). On the whole, 88% of Viscaceae and 61% Loranthaceae epiparasitize Loranthaceae (Calvin and Wilson, 2009).

Hyperparasitism in Other mistletoes in the same genus

Notothixos subaureus (Viscaceae), parasitic on Dendrophthoe glabrescens (Loranthaceae) on Eucalyptus (source: http://www.panoramio.com/photo/1876445)

Insects:

Recently, Cockerell Scale insects, Pseudaulacaspis cockerelli Cooley (Diaspididae) have been reported to settle on the leaves, stems and haustoria of D.falcata parasitizing Senna siamea (unpublished communication). These scale insects are known to thrive on sap nutrients and their reproduction and hence their proliferation rate has depicted relation with the nitrogen status of the host. Well in this regard mistletoe particularly D.F indicates a favorable choice as their nutrient uptake seems more biased towards drawing more nitrogen rich species from the host sap. The host S.siamea, though is a non-nodulating leguminous tree specie but can produce more nitrogen than the nodulating relatives.

Controlling the parasite:

Backed by easy seed dispersal by fruit eating birds, widespread and ever increasing host range of Dendrophthoe falcata continues to pose heavy losses in economically valuable fruit trees and those with medicinal properties whether growing in forests, orchards or gardens (Sridhar and Rama Rao, 1978).

Healthy tree hosts can tolerate few branch attachments with mistletoes but a heavily infected host slowly moves towards its decline as the mistletoe feeds upon its nutrients and spreads on all its ariel parts (Aly, 2007). Hosts¿½ death is more probable when they become more prone to other forms of biotic ad abiotic stresses.

Reports from 1930s evidence that until then no proper control and eradication methods were available rather the only alternative used was excising off the parasite (before seed production) completely from below the point of infection in the host and their immediate incineration (Sridhar and Rama Rao, 1978).

Mechanical pruning of the infected host branches with cuts at least one foot below the point of mistletoe attachment in order to completely remove embedded haustoria has also been suggested as a method of choice but severe heading (topping) which is often used to remove heavy tree infestations weakens the host tree structure. Instead, it is best to remove severely infested trees entirely because they are usually a source of mistletoe seeds. Another method employed when mechanical pruning gives up upon mistletoe infection uses painting the infected area using a pruning paint or spray and wrapping it with layers of opaque polyethylene cover so as to restrict light source required by this leafy mistletoe. In another method, use of ethephon (a growth regulator) in form of spray to completely wet leaves of mistletoe harboring dormant host trees is recommended (Aly, 2007). Regrowth of the parasite may be prevented by pruning it and painting the spot in the host branch (Koski et al., 2009).

Lately, adopted method of control uses application of 50% diesel oil emulsion sprayed on mistletoe leaves. The said morbid effect of emulsion significantly increases the intra-molecular respiration rates in the leaves followed by a gradual decline in capacity of stomatal closure consequently leading to death of the hemiparasite (Singh and Sharma., 1970).

Delias eucharis, a medium sized pierid butterfly (Common Jezebel) commonly occurs in South and South-East Asia (Bingham, 1907) and is particularly known to lay its eggs and feed on leaves of mistletoes (Roychoudhury, 2005) and has been suggested in its use to control the parasitic mistletoes (Sevastopulo, 1950, Mushtaque and Baloch 1979).

Alternatively, resistance to D.falcata using herbicide (metribuzin) infusion of its host, Tectona grandis (Teak) has also been tested but is of limited help as the efficiency of parasite killing was not proportional to the amount of herbicide used which is required to estimate exact dosage and infusion time for multiple of infection on one host (Balasundaran and Ali, 1989).

In one of the papers from the proceedings of the 6th Asian-Pacific Weed Science Society conference, G.G. Hambali (1977) has discussed beside other uses the exploitaton of scolytid and cerambycid beetles in controlling Dendrophthoe spp. Several similar or other control alternatives have been suggested by Khan et al. (1987).

Approaches as highlighted above towards control of parasitic weeds by traditional methods are limitedly effective and best control via herbicides and fumigants may be hazardous to Mother Nature (Jacobsohn 1994).

Removal of mistletoe by hand is effective considerably under small degrees of infection but extensive infestation might render this approach as inevitably expensive (Aly, 2007) and cumbersome more particularly when the infestation spans large tree plantations or dense forests.

Replacing the infected plantations with resistant trees has also been a method of choice. Though not reported yet specifically for D.falcata, resistance for broadleaf mistletoe (Phoradendron macrophyllum) was evidenced in some tree species viz., pear, Chinese pistachio, crapemyrtle, ginkgo, sycamore, and conifers such as redwood and cedar (Perry 1995). These few resistant species are of significance to the plant breeders but raising parasitic plant resistant crop genotypes have been very difficult and yet lack success.

Use of high and low intensity fires has also been noted but is rather undesirable in plantations as won¿½t result in complete cure but only subside heavy infestation from the mistletoes in stand without suppression of re-infestation (Carnegie et al, 2009).

Chemical control of mistletoes has been extensively researched in Australia which points out the reasons for their limited routine practice in forests as these methods are labour intensive, cost-effective, variable or limited in results. Use of biological control strategies including mistletoe targeting fungi has also been evidenced (Carnegie et al, 2009).

Recent efforts by S. Aly and his workgroup (at the Department of Plant Pathology and Weed Research, Agricultural Research Organization, Newe Yaar Research Center, Israel) enlists successful biotechnological approaches which guarantee prospects towards achieving effective parasitic weed control and ensure environmental safety (Aly, 2007). Adoption of these tools within biotechnology viz, trait gene introgression, gene silencing, promoter technology etc accordingly towards generating transgenic host trees remain the only promising alternative to combat losses incurred from D.falcata infestations.

Aspects of biodiversity:

Management and control strategies whether mild or exhaustive and whether practiced on a low to high area of coverage should not overlook that mistletoes are considered a valuable component of biodiversity. They are seen as ¿½a keystone resource¿½ of food and shelter for diverse population of vertebrates and invertebrates (Carnegie et al, 2009).

(KEY: take up the previous issues in above topics and be general and also take examples from other species or genera about the same, for medicinal values and products again search for other existing natural resources.

- Parasitic plants and their origin
- Parasitic plants constitute to about 1% of the total angiosperms in nature.
- Studies explaining the science behind the evolutionary origin of the ariel habit in Loranthaceae have recently been initiated where primitive hypotheses are being investigated with use of phylogenetic tools and analyses (Wilson and Calvin, 2006).

Phytochemistry: Though in a recent review by D'Cruz et al (2010) considerable reproductive toxicity that may cause upon consumption of extracts from the hemiparasite and other plant extracts had been advocated. The article also appeals for extensive research on neutralizing the toxicity of plant based products before they could be validated for their health benefits.

Medicinal uses: Antioxidants are considered important for preventing human diseases but the synthetic antioxidants viz., butylated hydroxy toluene (BHT) and butylated hydroxy anisole (BHA) that have a restricted use as preservatives in oils and fatty foods to minimize their oxidation are considered carcinogenic (Andlauer and Furst, 1998) which can be altered by use of ethanolic and/or methanolic D.falcata extracts as antioxidant supplements (Pattanayak and Sunita, 2008; Anarthe et al, 20082).

Hyperparasitism/epiparasitism

Despite being harmful for the economically important plants it is surprising though pertaining to their convincing medicinal properties that D.falcata is considered sacred (Kunwar et al.2005) in most if not all parts of India.

As reported by Pankaj Oudhia (2008), particularly in the state of Chhattisgarh (India) it is considered to be an integral part of traditional healing and the natives welcome its infestation and use it in hundreds of medicinal formulations. It is commonly known as Mahua ke Banda (friends of the host) when it occurs on the host Mahua (Madhuca indica) tree. Moreover, even the ethnobotanical surveys conducted in different parts of Indian state Chhattisgarh from year 1994 to 2008 revealed that different parts of Dendrophthoe falcata Ettingsh are used both internally as well as externally in treatment of many diseases and that information on over 1600 formulations in which it is added as main ingredient has been collected so far. In treatment of Type II Diabetes it is used by the traditional healers in combination with medicinal rice and medicinal Orchid species. Also reported from the native traditional healers¿½ viewpoint that this mistletoe takes initial support from trees and in general utilizes the waste or less important nutrient from these trees-they claim. Dendrophthoe falcata Ettingsh growing on different trees possess different medicinal properties. It is also claimed that Dendrophthoe falcata Ettingsh on Mahua trees growing in different locations possess different medicinal properties.

Damages incurred by D.falcata and other relatives

Static Bending Tests used in timber industries are regarded as predictors of ultimate strength and have successfully been performed on mistletoe infected Teak plantations (Gnanaharan et al, 1984). Standardizations for such tests on other healthy hosts and comparative analysis with infected hosts could provide a correct estimate of damage from mistletoes.

Other uses of the plant

A latest report have tested use of mucilage as tablet binders (Kothawade et al, 2010). In another recent attempt investigating repellent activity against red flour beetle, Tribolium castaneum Herbst (Coleoptera: Tenebrionidae) extract from D.falcata showed negligible success as compared to that from other epiphytes (Chhabi et al, 2010).

More studies required?

Pesticides are designed to kill, supposedly insects but not humans. Numerous pesticides are sprayed upon crops to keep them glamorous enough until they reach grocery stores and last until consumption. Even though assessing detailed health effects is still a scientific challenge, the accumulated health risks and detriments cannot be completely ignored. These toxins have been developed to eradicate insects but are causing side effects to our health too. The pesticide residues remaining on what we consume each day have proven to be harmful and can be attributed as the source of many untraceable and unexpected health problems in the long run.

While leading a more natural and healthy lifestyle has been under the spotlight in this age of artificialization, pesticide usage could be targeted as a root cause for many chronic diseases and prolonged health effects. The chemical composition of certain pesticides has been discovered to hinder the biochemical processes of vital organs

As soon as a pesticide enters the passageways of the body and comes in contact with any cell, it chemically reacts with its newfound surroundings, lowering the toxicity. These counter effects occur to make it more water soluble and easier to excrete out of the system. The body reacts to each pesticide depending on its chemical structure and properties including its shape, size, electronic charge, and how stable it is, defining how soluble the pesticide will be in different solutions and

mycelia remain on the surface of the host epidermis and are visible as white to grey spots. In rust diseases the pathogen becomes visible when the sporulating infections rupture the epidermis, exposing powdery masses of spores.

A large number of pathogens fall between these two types. The pathogen itself is not visible, but the symptoms of its presence are more or less easily discernible and restricted to the parts of the host tissue invaded. Discoloration of the invaded tissue and immediately adjacent tissue is the most general indication of the pathogen. The reliability of this measure of pathogen presence or inversely, host resistance, varies with host and pathogen but tends to be fairly good in many cases.

Inoculum Density

This factor may obscure real differences in quantitative resistance. In order to prevent escape of genotypes from infection, there is a tendency to apply high inoculum densities. Complete resistance in such cases is easily detectable, but small differences in susceptibility tend to disappear. The optimal inoculum density is
an
reac [illegible] density whereby escapes are largely prevented while only the most susceptible
as a byp [illegible] are strongly affected (Parlevliet, 1989).
caused by com [illegible]

Generally, any foreig [illegible] onsiderably in earliness the period of exposure to the set off asthma attacks as the airways [illegible] is usually done at the same moment for presence of pesticides in the respiratory system [illegible] aggressive response as the constriction of airways causes bronchiole contractions to occur, causing wheezing and breathlessness. The high level toxicity of pesticides that engulfs the lungs through residues from food puts your respiratory system under enormous stress, which is detrimental to victims of asthma.

Organophosphates are a type of phosphoric acid consisting of the elements phosphorus, carbon, and hydrogen. Especially seen on fruits we eat every day, organophosphates are repeatedly sprayed on pounds and pounds produce to keep them insect-free enough to catch your eye at the grocery store even though the EPA classifies them to moderate to highly toxic. Tests conducted by the USDA in 2008 found that '95% of celery tested contained pesticides, and 85% contained multiple pesticides, 93.6% of apples, 96% of peaches, a single blueberry has residue from 13 different chemicals,' (HuffPo).

The organic molecule, whose purpose is to function as a neurotoxin to insects has been now discovered as an inhibitor to multiple processes. Since cells aren't physically connected to one another, the synapses of neuron relay messages to the brain. These 'bridges' forming the entire network throughout the body are broken sorghi Schw., oats (Avena sativa L.) - P. coronata Cda., wheat [Triticum aestivum (L.) Thell.] - P. graminis f. sp. tritici Pers., wheat - P. recondita, wheat - P. striiformis Westend, barley - Erysiphe graminis D.C. f. sp. hordei, flax (Linum usitatissimun L.)-Melampsora lini (Ehrenb.) Desmaz. from over 20 to over 50 major resistance genes are known. These resistance genes often cluster together in certain chromosome arms, sometimes so tightly that they can be considered as complex loci, and true allelic series also occur. In the flax-flax rust pathosystem 34 R-genes have been identified in seven regions - K(2), L(14), M(7), N(3), P(6), D(1), Q(1). Regions N and P are linked, as well as regions N and K. The N region consists of at least two closely linked loci. The M region, together with seven resistance alleles, also consists of some closely linked loci. The L region, with 14 resistance alleles, behaves as a locus with an allelic series, but intra-allelic recombination has been reported (Islam and Shepherd, 1991). In barley most of the resistance genes to powdery mildew are located on one arm of chromosome 5 and one arm of chromosome 4 (Jorgensen, 1990). On the short arm of chromosome 10 in maize, at least 16 resistance genes to P. sorghi are found on the complex locus Rp1 and the loci Rp5 and Rp6 within three centimorgans of each other (Saxena and Hooker, 1968). The three downy mildew [Peronospora effusa (Grev.) Tul.] resistance genes known in spinach are tightly linked.

Minor or polygenic resistance has been reported fairly often. Typical examples are the quantitative resistance in maize to the northern [Setosphaeria turcica (Luttr.) Leonard & Suggs] and southern (Cochliobolus heterostrophus Drechsler) leaf blight (Jenkins and Robert, 1961; Burnette, and White, 1985) in barley to Puccinia hordei Otth. (Parlevliet, 1978), in wheat to P. recondita f. sp. tritici (Broers and Jacobs, 1989) and in rice to bacterial blight (Koch and Parlevliet, 1991). Minor genic or polygenic resistance is almost certainly ubiquitous as is discussed in the sub-chapter on quantitative resistance.

The expression of resistance genes can be modified by the action of other genes (epistasis), the development stage or tissue of the plant or the environment. The major resistance gene Pa7 in barley to P. hordei gives complete resistance in 'Cebada Capa', but incomplete resistance in the cultivars L94, Zephyr and Vada. In cereals, adult plant resistance genes to the various rusts occur frequently; they give resistance only in the adult plant stages. In the seedling stage the resistance is not expressed. In potatoes (Solanum tuberosum L.) the quantitative resistance in the foliage to late blight, Phytophthora infestans (Mont.) de Bary is poorly correlated with the resistance in the tubers, indicating that different genes are involved or expressed in the different tissues of the plant (Anonymous, 1958-1998). Of the environmental factors temperature plays a major role. Many scientists have reported that the expression of certain resistance genes depends on the temperature to which they are exposed, such as L1, L3, L7, L8, L10 and L11 in flax to flax rust (Islam and Mayo, 1990), Sr6 in wheat to stem rust, Lr16 and Lr17 in wheat to leaf rust and many other genes (Browder, 1985).

The Gene-for-Gene Concept

Many major resistance genes operate in a gene-for-gene way. For each resistance gene in the host there is a corresponding avirulence gene in the pathogen (Flor, 1971), and only the corresponding avirulence gene can initiate the hypersensitive reaction (HR) leading to incompatibility. Resistance and avirulence inherit in most cases in a dominant manner, susceptibility and virulence in a recessive way. The HR is now known to result from the specific interaction at the cellular level of the product of the resistance gene and the product of the avirulence gene. If one of the two products is absent, there is no incompatibility; the normal pathogenicity of the pathogen results in a compatible reaction (the host appears susceptible). What is normally meant with virulence is actually the normal pathogenicity shown in the absence of avirulence. Virulence is absence of avirulence, it is genetically seen as an empty concept; there are no virulence genes.

Host and pathogen genotypes interact with each other. Each of the four pathogen genotypes shows a different reaction pattern with the four host cultivars. This makes it QR is, except for a few cases of monogenic, incomplete inheritance, inherited oligogenically or polygenically. Examples of the former are the QR of maize to P. sorghi (Kim and Brewbaker, 1977) and of wheat to P. recondita possible to differentiate between the four pathogen genotypes (races), but only because the interaction effects, which are equal to the gene effects, are so large (difference between 80% and 0%).

7

BIOLOGICAL CONTROL

Biological control can be defined as the use of an organism to reduce the population density of another organism and thus includes the control of animals, weeds and diseases. In this article, we focus on the biological control of arthropods, which DeBach (1964) defined as 'the study and uses of parasites, predators and pathogens for the regulation of host (pest) densities'. This definition establishes two of the main principles of biological control. Firstly, in nature, most organisms are consumed by other organisms, which in many cases leads to drastic reductions in the population of the prey species; in biological control, man exploits this 'natural control' to suppress the numbers of pest species. Secondly, biological control reduces rather than eradicates pests, such that the pest and natural enemy remain in the agro-ecosystem at low densities. A number of important pests can be kept at a low population density by biological control over long periods of time. In other cases, populations of pests are significantly reduced by natural enemies, but repeated releases or additional methods are needed to achieve an adequate level of control. These methods include, for example, resistant plants, cultural techniques, physical barriers, semiochemicals and, as a last resort, the use of selective chemicals; this is the fundamental philosophy of integrated pest management (IPM; Stern et al. 1959).

Many biological control schemes use predatory insects and mites, insects that parasitize other insects (parasitoids) or nematodes, targeted against insect and mite pests; these are the so-called 'macrobial' agents. There are also various 'microbial' agents (bacteria, viruses and fungi) that have been developed and applied in arthropod biological control. Herbivorous insects and mites have also been used in the biological control of weeds (Bellows & Fisher 1999; van Lenteren 2003). Biological control schemes operate throughout the world as part of the management of pests in agriculture, forestry and greenhouse horticulture. Although biological control has sometimes been introduced to combat arthropod pests that have developed resistance to insecticides and acaricides, the first biological control programmes pre-date the modern pesticide era. The intensification of agriculture through the twentieth century has been accompanied by increased international trade and the resultant transfer of pest species on plants and products across the globe. In addition, the introduction of new crops (such as soya in central Europe), and consumer demand for 'blemish-free' produce, have all contributed to an over-reliance on chemical control, with the well-described consequences of pest resistance, uneconomic production costs, bioaccumulation through food chains, environmental pollution, loss of biodiversity and risks for human health.

At the start of the twenty-first century, the opportunities and need for effective biological control are greater than ever: pest resistance continues to be a problem, pesticides are being withdrawn on environmental grounds without suitable replacements, and in some areas of the world, including Europe, there is uncertainty over the durability and public acceptance of genetically modified (GM) pest-resistant crops. Against this background, this paper defines the terms used to describe different types of biological control and reviews landmark events in the development of this approach. The advantages and limitations of using natural enemy species in pest management are evaluated, as well as the common features of the most successful agents. The contribution of biological control to sustainable agriculture is reviewed on a global scale, together with recent information on the possible effects of GM crops on natural enemies of pests, and developments in the regulatory framework and risk assessment for the import and release of non-native species. We will explain how biological control can contribute to sustainable agriculture and increase biodiversity.

Natural enemies of insect pests, also known as biological control agents, include predators, parasitoids, pathogens, and competitors. Biological control agents of plant diseases are most often referred to as antagonists. Biological control agents of weeds include seed predators, herbivoresand plant pathogens.

Biological control can have side-effects on biodiversity through attacks on non-target species by any of the same mechanisms, especially when a species is introduced without thorough understanding of the possible consequences.

HISTORY

The term "biological control" was first used by Harry Scott Smith at the 1919 meeting of the Pacific Slope Branch of the American Association of Economic Entomologists, in Riverside, California. It was brought into more widespread use by the entomologist Paul H. DeBach (1914–1993) who worked on citrus crop pests throughout his life. However, the practice has previously been used for centuries. The first report of the use of an insect species to control an insect pest comes from "Nan Fang Cao Mu Zhuang" (WS[1]eIƒ(gÀr Plants of the Southern Regions) (ca. 304 AD), attributed to Western Jin dynasty botanist Ji Han (G]+T, 263–307), in which it is mentioned that "Jiaozhi people sell ants and their nests attached to twigs looking like thin cotton envelopes, the reddish-yellow ant being larger than normal. Without such ants, southern citrus fruits will be severely insect-damaged". The ants used are known as huang gan (huang = yellow, gan = citrus) ants (Oecophylla smaragdina). The practice was later reported by Ling Biao Lu Yi (late Tang Dynasty or Early Five Dynasties), in Ji Le Pian by Zhuang Jisu (Southern Song Dynasty), in the Book of Tree Planting by Yu Zhen Mu (Ming Dynasty), in the book Guangdong Xing Yu (17th century), Lingnan by Wu Zhen Fang (Qing Dynasty), in Nanyue Miscellanies by Li Diao Yuan, and others.

Biological control techniques as we know them today started to emerge in the 1870s. During this decade, in the USA, the Missouri State Entomologist C. V. Riley and the Illinois State Entomologist W. LeBaron began within-state redistribution of parasitoids to control crop pests. The first international shipment of an insect as biological control agent was made by Charles V. Riley in 1873, shipping to France the predatory mites Tyroglyphus phylloxera to help fight the grapevine phylloxera (Daktulosphaira vitifoliae) that was destroying grapevines in France. The United States Department of Agriculture (USDA) initiated research in classical biological control following the establishment of the Division of Entomology in 1881, with C. V. Riley as Chief. The first importation of a parasitoidal wasp into the United States was that of the braconid Cotesia glomerata in 1883–1884, imported from Europe to control the invasive cabbage white butterfly, Pieris rapae. In 1888–1889 the vedalia beetle, Rodolia cardinalis, a lady beetle, was introduced from Australia to California to control the cottony cushion scale, Icerya purchasi. This had become a major problem for the newly developed citrus industry in California, but by the end of 1889 the cottony cushion scale population had already declined. This great success led to further introductions of beneficial insects into the USA. In 1905 the USDA initiated its first large-scale biological control program, sending entomologists to Europe and Japan to look for natural enemies of the gypsy moth, Lymantria dispar dispar, and brown-tail moth, Euproctis chrysorrhoea, invasive pests of trees and shrubs. As a result, nine parasitoids (solitary wasps) of gypsy moth, seven of brown-tail moth, and two predators of both moths became established in the USA. Although the gypsy moth was not fully controlled by these natural enemies, the frequency, duration, and severity of its outbreaks were reduced and the program was regarded as successful. This program also led to the development of many concepts, principles, and procedures for the implementation of biological control programs.

Prickly pear cacti were introduced into Queensland, Australia as ornamental plants, starting in 1788. They quickly spread to cover over 25 million hectares of Australia by 1920, increasing by 1 million hectares per year. Digging, burning and crushing all proved ineffective. By 1914, two control agents were introduced to help control the spread of the plant, the cactus moth Cactoblastis cactorum, and the scale insect Dactylopius. By 1917, most areas of prickly pear had been destroyed.

The first reported case of a classical biological control attempt in Canada involves the parasitoidal wasp Trichogramma minutum. Individuals were caught in New York State and released in Ontario gardens in 1882 by William Saunders, trained chemist and first Director of the Dominion Experimental Farms, for controlling the invasive currantworm Nematus ribesii. Between 1884 and 1908, the first Dominion Entomologist, James Fletcher, continued introductions of other parasitoids and pathogens for the control of pests in Canada

Types of biological control

There are different types of biological control, but both the exact number of types, and the definition of each type vary depending on the literature consulted. The situation is further complicated when some terms are used synonymously by one author, for example 'classical biological control' and 'introductions' by Dent (2000), and the 'same' type of control (classical) is described by a different synonymous term ('inoculation') by another author (van Emden 1989), with the first author also regarding introductions and inoculation as being distinct and different.

In the context of this article, biological control will firstly be distinguished from natural control. Thus, biological control is the use of an organism by man to reduce the population density of another organism, whereas natural control is the reduction in numbers of the population of a species by a naturally occurring natural enemy with no human intervention.

There are three main techniques of biological control: classical (sometimes described as 'inoculative' biological control), augmentative (where a distinction can be made between 'inundation' and 'seasonal inoculation') and conservation control (van Lenteren 1993a, 2000b). Classical or inoculative control is used mainly against 'exotic' pests that have become established in new countries or regions of the world. Relatively, small numbers (usually less than 1000) of a certain species of natural enemy are collected from the country or region of origin of the pest, 'inoculated' into the new environment, and allowed to build up the level of control, which can be maintained over very long periods of time. This type of biological control has been most successful with perennial crops (fruit plantations and forests), where the long-term nature of the ecosystem enables the interactions between pest and natural enemy to become fully established over a period of time: for example, the successful import and release of the predatory ladybird Rodolia cardinalis for control of the accidentally introduced citrus pest Icerya purchasiin Mediterranean Europe around 1900, and release of the parasitoid Aphelinus mali in the 1930s for control of the accidentally imported woolly apple aphid, Eriosoma lanigerum, in apple orchards throughout Europe (Greathead 1976). For a comprehensive review of examples of classical biological control, see DeBach (1964) and Bellows & Fisher (1999).

Augmentation refers to all forms of biological control in which natural enemies are periodically introduced, and usually requires the commercial production of the released agents (van Lenteren 2000b). Inundation involves the mass production and release of large numbers of the control agent, such as the Trichogramma egg parasitoids of various lepidopteran pests including the cotton bollworm Heliothis virescens, the sugar cane borer Diatraea saccharalis and the European corn borer Ostrinia nubilalis (Bigler 1986; Li 1994; Smith 1996; van Lenteren & Bueno 2003). The cultivation systems used for short-term annual crops destabilize the pest–natural enemy relationship, and thus often prevent the establishment of viable breeding populations of natural enemy species between the crop production cycles. The aim of inundative releases is to create a massive ratio in favour of the natural enemy, analogous to the use of a pesticide, producing a rapid reduction or local extinction of the pest. Control is achieved mainly by the individuals that have been released rather than their offspring. However, the control is usually transient, and re-releases are required, sometimes more than once a year.

Seasonal inoculative control is a form of augmentation in which natural enemies are similarly mass reared in the laboratory and periodically released into short-term crops where many pest generations can occur in each growing season (van Lenteren & Woets 1988). As with augmentative control, relatively large numbers of natural enemies are released to obtain immediate control, but in addition, a build-up of the natural enemy population occurs through successive generations during the same growing season. This method can be applied when the cropping system prevents control extending over many years as with classical (inoculative) biological control. Examples include the control of whiteflies, leafminers, thrips, aphids and mites by parasitoids and predators in greenhouses (van Lenteren 2000a).

Conservation control refers to the use of indigenous predators and parasitoids, usually against native pests. Various measures are implemented to enhance the abundance or activity of the natural enemies, including manipulation of the crop microclimate, creation of overwintering refuges (like 'beetle banks'), increasing the availability of alternative hosts and prey, and providing essential food resources such as flowers for adult parasitoids and aphidophagous hoverflies (Gurr et al. 2000; Wäckers 2003; Winkler et al. 2005).

2. Landmark Events In Biological Control

The first historical record of biological control dates back to around AD 300 when predatory ants were used to control pests in citrus orchards (van Lenteren 2005; van Lenteren & Godfray 2005). In the modern era, the control of the cottony-cushion scale I. purchasi on citrus crops in California in the 1880s by an imported ladybird (the vedalia beetle R. cardinalis) and dipteran parasitoid (Cryptochaetum iceryae), is widely regarded as representing the first major success of classical biological control (DeBach 1964). The scale insect was accidentally introduced to the United States on Acaciaplants imported from Australia. Two natural enemy species of I. purchasi were collected in Australia, released in small numbers in Californian citrus groves and achieved control of the scale in less than 2 years, the ladybird being the most effective agent. The success of this scheme, which has been repeated in other locations against the same pest worldwide, is attributable to a number of ecophysiological characteristics of the predator–prey relationship: both the adult and larval stages of R. cardinalis were active and could seek out their sedentary prey, the rate of development of the ladybird was more rapid than the scale, and the predator had no natural enemies of its own.

Another early example of classical control involved the coconut moth Levuana iridescens in Fiji, and established other important principles in the development of biological control. A tachinid parasitoid Ptychomyia remota was known to attack the moth Artona catoxantha in Malaysia. The parasitoid was introduced to Fiji to control a 'new' host which it had not previously encountered (L. iridescens), and was successful across the whole island within 2 years. The stable Fijian climate allowed year-round reproduction of L. iridescens, providing a continuous source of prey for P. remota; additionally, the island ecosystem was small and isolated (DeBach & Rosen 1991).

One of the most recent successes in biological control is a further example of classical control. Cassava (Manihot esculenta) is one of the staple diet crops for a number of African countries. Cassava was introduced to Africa from South America and had been grown for hundreds of years without any major pest problems, until the 1970s. The crop then became infested with an accidentally imported mealy bug from South America Phenacoccus manihoti, which spread rapidly over the next 15 years through many countries. A search in South America found the pest in Paraguay, together with one of its natural enemies, the parasitoid Epidinocarsis lopezi. The parasitoid showed good searching ability, apparently locating its host by attraction to damaged leaves, and was able to develop more quickly than the mealy bug. After its first release in Nigeria in 1981, E. lopezi spread rapidly through neighbouring African countries and is now regarded as one the most successful programmes in biological control with enormous economic benefits (Neuenschwander et al. 2003).

One of the first successes with inundative control on a field scale involved the citrophilus mealy bug, Pseudococcus calceolariae Fernald (equal to gahani Green), a pest of citrus in southern California (Luck & Forster 2003). With the discovery and spread of the mealy bug in 1913, the citrus growers were faced with a serious problem that threatened the crop's viability. In response to this problem, the citrus industry established insectaries to mass-produce the predatory beetle Cryptolaemus montrouzieri(Smith & Armitage 1920, 1931; Quayle 1938). Mass production techniques were developed by the early 1920s and, for the next decade, the beetle was the principal means of suppressing this pest. In 1930, at the height of the P. calceolariae campaign, 16 insectaries had been established and were producing 20 million beetles annually for release in infested groves (Luck & Forster 2003).

Inundative releases are currently used on a large scale against a range of pests. In Brazil, for example, Cotesia parasitoids are applied against sugar cane borer on 300 000 ha, Anticarsia gemmatalis multiple nucleopolyhedrovirus (AgMNPV) against soya bean caterpillar on more than 1 Mha, and egg parasitoids of soya bean bugs on 20 000 ha. Egg parasitoids of the genus Trichogramma are now released on more than 10 Mha worldwide (van Lenteren & Bueno 2003).

The first success of seasonal inoculative biological control in protected cultivation involved the glasshouse whitefly Trialeurodes vaporariorum and the parasitic wasp Encarsia formosa. The whitefly and other greenhouse pests (spider mite, thrips and aphids) attack a range of vegetable crops (tomato, cucumber and peppers) and ornamental flowers. These pests reduce yields, lower the market value of products that are sold on appearance, and provide a breeding medium for secondary infections such as the sooty mould (Cladosporium sphaerospermum). Biological control of T. vaporariorum was established before 1930 in the UK and ran successfully until the introduction of DDT and other organochlorine insecticides in the 1940s (van Lenteren & Woets 1988). However, the development of resistance in the whitefly, together with other concerns about the overuse of pesticides led to the re-introduction of the biological control scheme in the 1970s. For both T. vaporariorum and the glasshouse spider mite Tetranychus urticae controlled by the predatory mite Phytoseiulus persimilis, sophisticated management schemes have now been developed, involving the commercial production of the control agents, flexible release plans (frequency and density), introduction of the pest species to maintain natural enemy populations when required, and manipulation of the climatic environment (temperature and humidity) to optimize control. Currently, all greenhouse vegetable pests can be managed with biological control agents (van Lenteren 2000a).

A major success in conservation biological control was achieved with pest management in European fruit orchards. A few years after the use of blanket sprays of synthetic pesticides there were marked declines in the abundance of several species of naturally occurring predators and parasitoids. A series of related studies investigated the occurrence of pests and natural enemies in sprayed and unsprayed orchards, and the relative toxicity of different pesticides on the beneficial species. This information was used to design an IPM programme based on the re-establishment of the natural enemy fauna (predatory insects, mites and parasitoids) to control key pests such as leaf rollers, leaf miners and spider mites, and the selective and local use of chemicals against other pest species (Gruys 1982).

(a) Advantages and limitations of biological control

The advantages and limitations of biological control are often expressed by comparisons with pesticides. Thus, predators and parasitoids are naturally occurring organisms and usually fairly specific in the range of prey that they will attack. Natural enemies actively seek out their prey and can increase the level of control over time. It is unlikely that resistance will develop to a control agent, and in many cases, the control can be self-perpetuating over long periods of time. The arguments against chemical pesticides are that they not only kill the pest organism, but also many non-target species, including natural enemy species, which in turn, may increase the pest status of species that were previously unimportant or easy to control. Furthermore, chemical control is limited to the area within which the pesticide is applied, frequent application may be required, and this selects for pest resistance.

The main limitation of biological control is that it is slower to suppress pest populations than most pesticides as parasitized organisms may take several days to die; and also, predators require a period of time to establish an economic level of pest suppression. Development costs of biological control are sometimes described as 'high', but these costs are much lower than for the equivalent synthesis, toxicological evaluation and marketing of a new pesticide, and substantial profits can be achieved from biocontrol with long term, effective natural enemies, of which R. cardinalis, E. lopezi, E. formosaand P. persimilis are good examples. The comparative advantages of pesticides are that control is rapid, and in the absence of resistance, a high and predictable level of mortality is normally guaranteed.

There are other reported limitations of biological control where pesticides seem to be the more advantageous approach, but the evidence for this conclusion is limited or absent (van Lenteren 1993a,b). For instance, the fact that biological control 'regulates' whereas pesticides 'eradicate' is viewed as disadvantageous, but this is a spurious argument. In biological control, pest populations are usually not eradicated, but maintained at very low densities; long-term suppression of pest species is the desirable aim. Eradication of a pest population after chemical treatment, even if this is achieved, occurs only on local scales, and the environment is then open to re-invasion, often with a much reduced natural enemy fauna.

Biological control has been described as 'unreliable' when compared with pesticides, but again the information for this view is equivocal. Some biological control schemes have been 'partial' rather than 'complete' successes, and in some instances, there have been periodic fluctuations in the level of control (see DeBach (1964) and Bellows & Fisher (1999) for examples). But as a counter argument, pesticides also vary in their effectiveness and over time, the development of resistance can lead to the failure of a previously successful chemical.

It is now widely acknowledged that successful biological control depends on extensive preliminary studies to gain a comprehensive understanding of the biology and ecology of the pest and natural enemy complex, and of the environments from which they originate and have subsequently colonized, or into which they will be released. This research can be time consuming, but it is essential, and analogous to the period of time (estimated to be around 10 years) between the synthesis of a new molecule as a potential pesticide and placing a product on the market. Where all of the necessary 'efficacy studies' have been carried out, such as with P. persimilis and E. formosa, the control programmes have been consistently successful. In other situations, and under strong market pressure, control agents have sometimes been released without adequate evaluation, and control has been less successful and the pest problems have continued; for example, predatory mites released against Frankliniella occidentalis.

It has been suggested that the introduction of a biological control scheme against a pest, and the resultant withdrawal of broad-spectrum pesticides, can lead to new pest problems, but there is no evidence for this in several well-studied agro-ecosystems, including glasshouse systems. Research on biological control was undertaken from 1965 to 1975 to combat glasshouse pests (P. persimilis and E. formosa) that had become resistant to the then available pesticides (DDT, malathion and resmethrin), but the wide-scale introduction of natural enemies against these two species did not by itself lead to the occurrence of any new pests. The pest problems that have arisen in the European glasshouse industry since 1975 (Spodoptera exigua, Liriomyza trifolii, Liriomyza huidobrensis, F. occidentalis and Bemisia tabaci) have all been unintentional introductions that arrived with varying levels of resistance to most pesticides. These newly imported pests have threatened the biological control of existing pests because governments usually respond to pest invasions by initiating extermination programmes based on frequent application of pesticides, thereby killing the natural enemies of the 'old' pests. Therefore, it is crucial to identify effective natural enemies of new pests before or soon after invasion, in order to maintain stability in the commercial biological control of existing pests.

Cost–benefit analyses suggest that research on biological control is more cost effective than on chemical control (30 : 1 and 5 : 1, respectively; Tisdell 1990; van Driesche & Bellows 1995). Despite this advantage, the main reason that biological control is not used on a larger scale is attributable to problems associated with the production and distribution of parasitoids and predators; particularly, the limited shelf life (days or at most weeks) of most natural enemy species and the impossibility of patenting a naturally occurring, unmodified species.

It is evident that many more chemical compounds have been tested than species of natural enemy (table 1), though there are still vast numbers of predator and parasitoid species that remain to be screened for their use in biological control. The success rate for finding a successful natural enemy is much higher than for a chemical compound, which is mainly attributable to the 'directed search' that is used for natural enemies compared with the more random approach for chemical compounds, although pesticide discovery and design has become far more rational over time. Development costs are much higher for chemical pesticides, largely as a result of the very stringent requirements for ecotoxicological studies as part of the registration process. Interestingly, the developmental time of an effective product, be it a chemical pesticide or a natural enemy species, is the same. It is common for insect and mite pests to develop resistance against chemical pesticides, whereas as resistance against natural enemies is unknown, at least to the extent that the control agent becomes ineffective. When compared on the criterion of specificity, even the most selective chemical is likely to kill many species of arthropod, whereas natural enemy species used in biological control are usually highly specific, killing only one or a few related species of prey. When selective natural enemies are used in biological control programmes, harmful side effects do not occur, but with highly polyphagous species there may be some negative environmental effects (van Lenteren et al. 2006a).

The favourable economics of biological control relative to the use of pesticides also applies to the commercial application of the technique. In the 1980s, Ramakers (1982)estimated the cost (product and labour) of controlling glasshouse whitefly (T. vaporariorum) to be twice as expensive using insecticide compared with the parasitoid E. formosa. A similar difference in costs was found with the control of spider mites (T. urticae) by predatory mites compared with acaricides (van Lenteren 1990). In the control of tomato and cucumber pests in the UK, Wardlow (1992) estimates chemical control to be three- or five times more expensive than using commercially produced natural enemies, and even where different biological control agents are used in combination (as with cucumber), the costs are still not higher than when using chemicals (Ramakers 1982).

Periodic analyses of successes and failures of biological control have suggested a number of more generalized limitations that relate to the types of pests that can be more easily controlled by predators and parasitoids, features of the targeted ecosystem and suitability of climates. For classical biological control, some of the most effective schemes have been applied against relatively sedentary pests (such as scale insects and mealy bugs), living in island ecosystems and in year-round stable climates that allow continuous development and reproduction of the pest and natural enemy (DeBach 1964). This view gained support from the control of I. purchasi in the citrus groves ('ecological islands') of California, L. iridescens in Fiji and sugar cane pests in Hawaii. However, as the scientific understanding of biological control has increased, through both theoretical and practical studies, it is apparent that these limitations are not an insurmountable obstacle to success. For example, biological control schemes have been successfully applied against mobile pests such as winter moth (Operophtera brumata) in Canada (by the parasitoids Cyzenis albicans and Agrypon flaveolatum from Europe), where there is a strongly seasonal climate with severe winters (DeBach & Rosen 1991). Nevertheless, it is evident that the seasonal temperate climate can be a barrier to the long-term survival of natural enemies originating from tropical or semi-tropical climates, such as the failure of the parasitoid A. mali to control the woolly aphid E. lanigerum in UK apple orchards in the 1920s and 1930s, through a lack of overwintering ability, even though the natural enemy had been successfully introduced into other countries. However, the main reason for the limited success of biological control in the field in northern Europe is not the weather, but the major changes that occurred in agricultural production systems after 1945 with the introduction of pesticide-dominated methods of pest control, and the loss of resistance to pests and diseases through plant breeding programmes that focused selectively on yield (Lewis et al. 1997). Where IPM schemes have been established in orchards, and thus pesticide use is highly restricted (or terminated),

Selection of successful agents

In the early stages of development, biological control of necessity followed mainly a 'trial and error' approach. But now, with a history of over 100 years, there is a large and increasing database of information that can be analysed with the aim of identifying the causes of success and failure, and hence the features that are commonly associated with the most successful agents. These analyses suggest that successful biocontrol agents show strong searching ability to locate prey and identify areas of high pest density, and ideally should have higher potential rates of increase than their prey (faster development, more generations per year, and greater fecundity or a greater predation rate than the reproduction capacity of the pest). Control agents must be able to survive at low pest density, otherwise control will fail and repeated releases will be necessary. The target pest should be the preferred prey of the natural enemy, though use of alternative prey is desirable during periods when the pest occurs at a very low density, or is not in a stage of the life cycle that is parasitized or predated by the control agent. However, it is vitally important that the released agent should not attack non-target species, and this feature is a key element in the pre-licensing environment risk assessment that is now routine in many countries. In situations where the natural enemy is strongly host specific, it is important for the life cycles of the predator and prey or parasitoid and host to be synchronous, especially in seasonal climates where diapause and the timing of post-winter emergence can influence the respective phenologies from year to year. The ability of parasitoids to detect and avoid hosts that have already been parasitized is also an advantageous feature of successful agents. The similarity in climate between collection and release sites is a further important consideration that can have a major impact on successful establishment in new environments.

It is clear that such analyses are not a guarantee of success, and there are also contradictory views among scientists that question some of the fundamental theories of biological control upon which such analyses are based. For instance, it has been argued that the most desirable biological characteristics identified in this selection process are derived from theoretical models of the interactions between prey species and predators or parasitoids without any strong evidence that such theory actually increases the frequency of success or provides an explanation for failures (Gutierrez et al. 1994). Likewise, it has been suggested that parasitoids and predators co-evolve with their prey such that control agents would be expected to become less effective over time (Pimentel 1963). If true, this suggests that the concept of classical biological control in which exotic pests are 'reunited' with their former natural enemies collected from the country or area of origin of both species is likely to be successful for limited periods of time; and conversely, that species collected from different locations to the native area of the pest, or from other host or prey associations, may be more successful, or maintain control for longer periods of time (Hokkanen & Pimentel 1984). Analysis of databases of successes and failures in biological control have not produced any strong evidence to support these ideas (Waage & Greathead 1988), and some of the classical control schemes based on the collection of natural enemies associated with the pest from its country of origin (cottony-cushion scale in California and sugar cane leafhopper in Hawaii) have now been operating successfully for 50–100 years. It also has to be recognized that there are important differences in predator–prey relationships in natural environments and those occurring with pests and biological control agents in agricultural ecosystems, particularly in terms of the density and uniformity of the plant hosts; as such, observations and predictions from naturally co-evolved interactions may not be a reliable predictor of outcomes in cropping regimes.

(c) Increasing the success of biological control

Some biological control schemes have the potential to be successful but require additional inputs from man, either prior to release or during natural colonization of the crop environment or to encourage post-release establishment, and this is particularly true of conservation biological control. As examples, adult parasitoids and hoverflies require sources of nectar and pollen to mature their eggs (Wäckers 2003), and growers have been encouraged to plant selected nectar-producing flowers around field boundaries to increase the attractiveness of the crop environment to such natural enemies, with some experimental evidence that an increase in crop diversity with flowers can increase the natural control of pests (Landis et al. 2000; Winkler et al. 2005).

An increase in plant diversity can also attract insect herbivores that act as alternative hosts or prey for the released agents. The parasitoid Anagrus epos parasitizes the eggs of grape leafhoppers (Erythromeura species) in California. The grape leafhopper overwinters in the adult stage, leading to a loss of synchrony in the host–parasitoid relationship. A different leafhopper (Edwardsiana prunicola) associated with prune trees provides an alternative host for A. epos, ensuring effective control of the grape leafhoppers at the start of each growing season (Pickett et al. 1990). The importance of alternative prey has also been identified in the control of western flower thrips (F. occidentalis) by the anthocorid bug Orius majusculus (Brodsgaard & Enkegaard 1997) and with the egg parasitoids of rice plant hoppers (Yu et al. 1996).

Sometimes, alternative prey, together with natural enemies, can be offered on non-crop plants adjacent to the crop, to enhance natural enemy populations by the so-called 'banker plant system'. As an example, one of the most effective methods of suppressing aphid populations in protective cultivation is to introduce the natural enemy species into the greenhouse even before any aphids have been discovered on the crop plants. This can be achieved by using open rearing units ('banker plants') consisting of wheat plants with wheat aphids (which cannot develop on the greenhouse crops) together with predators or parasitoids (van Steenis 1995). In other situations, slow-release systems are used to obtain a regular re-infestation of the crop with natural enemies as with the 'sachet-system' used for the control of thrips by predatory mites. Each sachet contains the predator with an alternative prey source; the mites consume the provided prey and emerge from the sachet progressively over several weeks and then predate the target pest (K. Bolckmans 2006, personal communication).

Assistance from man may extend to more specialized release systems such as spray equipment and machinery that 'shoots' small containers loaded with natural enemies into the crop (Trichogramma in maize), release by model airplane (whitefly parasitoids in infested patches of cotton), and full field release by helicopter, ultralight or small airplane (release of Trichogramma in maize and other crops). One of the most important actions man can take is to aid the synchronous occurrence of pest and natural enemy by well-timed releases of the control agent. A full review of this area is given by van Lenteren (1987) and Gurr et al. (2000).

3. CURRENT USE OF BIOLOGICAL CONTROL

Biological control can be defined as the use of an organism to reduce the population density of another organism. Natural biological control ensures that the Earth is 'green' and that plants can produce sufficient biomass to sustain other forms of life. Without biological control, the production of energy by plants would be a small fraction of current levels.

Natural biological control of pest organisms has occurred since the evolution of the first ecosystem some 500 Myr ago, and continues to the present day across 55.5 billion hectares of the world's terrestrial ecosystems and without human intervention. Its role in marine or freshwater ecosystems has not been quantified, but is certainly equally important as in terrestrial environments. In forest and agro-ecosystems, most of the potential arthropod pests (95% of the total or approximately 100 000 species; DeBach & Rosen 1991; J. C. van Lenteren 2000, unpublished data) are under natural biological control. All other modern methods of control are targeted at the remaining 5000 arthropod pest species. What is often unknown or easily ignored is that the maintenance of ecosystem functions achieved by natural biological control is estimated to have a minimum value of 400 billion US$ per year (Costanza et al. 1997), which is an enormous sum of money compared with the annual expenditure on insecticides of only 8.5 billion US$.

Classical biological control is estimated to be applied on 350 Mha worldwide (10% of land under culture; J. C. van Lenteren 2005, unpublished) and has very favourable cost–benefit ratios of 1 : 20–500. As soon as a natural enemy has been released and becomes effective, it contributes annually to benefits, while costs are minimal. As with natural biological control, the continuously increasing benefits of classical biological control programmes are often overlooked once a pest disappears as a result of the release of an efficient natural enemy. The comparative estimate for area usage of augmentative biological control is 16 Mha (0.046% of land under culture

Before the large-scale application of chemical pesticides, biological control was one of the main pest management methods embedded in a 'systems approach' to pest prevention and reduction, covering animals, weeds and pathogens. A farmer had to think about pest prevention before designing the next season's planting scheme and choice of crops. This concept generally made use of three methods of pest management: cultural control, host-plant resistance and biological control. Cultural methods such as crop rotation, cover crops and manipulation of sowing and harvesting dates were used to prevent the build-up of pest numbers (Delucchi 1987). Plants that had a high degree of resistance or tolerance to pests were another cornerstone of pest prevention, and the third was natural, classical, inundative and conservation biological control.

From around 1950, these methods became redundant as almost all pests could be easily managed by the newly discovered pesticides. As a result, pest control research became a highly reductionist activity, and changed from an important decision-making exercise in pest management to a routine but initially successful fire brigade activity. Another effect was that plants were no longer selected for resistance to pests, but only for the highest production of biomass (food) or cosmetic appearance (flowers), produced under blanket applications of pesticides. This, in turn, has led to the current situation in which crops have become 'incubator plants', unable to survive without frequent pesticide applications and agro-ecosystems that have few, if any, natural enemy populations.

However, now that chemical pesticides are no longer seen as the major solution for sustainable pest management, it is not possible to simply return in a year or so to pre-pesticide methods of control: crops are too weak to survive without pesticides, natural enemy faunas are absent, and many farmers remain 'pesticide-addicted'. So, first there is a need to invest in the development of new crop cultivars with resistance to pests and diseases, and there is evidence that this is happening in some countries. For example, the Dutch plant breeding industry now invests 35% of its research funds in the development of pest resistance compared with only 5%, 20 years ago (J. C. van Lenteren 2005, unpublished). At the same time, there is a need to restore previously used natural, classical, inundative and conservation biological control (e.g. control of spider mites and insect pests in apple orchards in several European countries; table 2 in van Lenteren 1993b). Further, several other alternatives to conventional chemical pest control can also be implemented such as mechanical, physical, genetical, pheromonal and semiochemical control. Also, the environment can be manipulated to make it more attractive and favourable for natural enemies. This strategy involves a range of methods, from manipulation of biotic and abiotic components of the environment such as modifying the climate (e.g. greenhouses and wind shields), to the application of chemicals that stimulate the activity of natural enemies. If natural enemies fail to establish (either due to agricultural practices or lack of adaptive ability in the natural enemy) or, if once established, fail to control the pest, manipulation of the natural enemy or its environment may lead to greater control. The local habitat may lack certain key requisites, which when provided may allow natural enemies to establish or become more effective. Manipulation of the environment is currently applied on a limited scale though there are many opportunities for implementation (see van Lenteren (1987) and Landis et al. (2000) for reviews).

The successful introduction (or more accurately, 're-introduction') of these 'new' pest management strategies will require retraining of the extension and advisory service and of growers. This is easier said than done, because it is difficult to simply replace a certain pesticide with an alternative method of control. Instead, there is a need to return to the systems approach, where the influence of all factors affecting pest abundance is considered. An example of a successful systems approach is seen with the optimization of fertilizer use, where a considerable reduction in fertilizer application reduced the development of several pests including aphids, whiteflies and leaf miners. The aim of this approach is to create a system that is inherently resistant to many pests and, thus, requires fewer or no treatment with conventional pesticides. Here, it is important to recognize that maximizing net income is not synonymous with maximizing yield. High yields are obtained with excessively high inputs of fertilizers and pesticides. Reducing the level of inputs may lead to somewhat lower yields, but as the financial inputs are also lower, the net income may be the same or higher. In general, integrated farming takes a more complete account of the impacts of agriculture on ecosystems (pesticide and fertilizer pollution of soil and water, preservation of flora and fauna, quality and diversity of landscape, conservation of energy and non-renewable resources), as well as sociological considerations (employment, public health and well-being of persons associated with agriculture), than is the case with conventional farming practices (Vereijken et al. 1986; Wijnands & Kroonen-Backbier 1993).

Although research on integrated farming is still limited, this approach is gaining impetus in Europe. The practices which can be manipulated in integrated farming programmes are crop rotation, cultivation, fertilization, pesticide use, cultural control measures, biological control and other alternatives to conventional chemical control. The practical results obtained in a large, long-term project in The Netherlands indicate that environmental pollution can be reduced through a decrease in fertilizer use and the replacement of chemical pesticides by non-chemical measures. In integrated farming, artificial fertilizers tend to be replaced by organic manure, reducing the total amount of freely available nitrogen (N) in the system, and hence the crop is less attractive or suitable for pest and disease development. Weed, pest and disease problems are reduced in integrated farming through the use of weed-competitive or disease- and pest-resistant varieties, reduction of N-fertilization, adoption of specific sowing dates and plant spacing, mechanical weed control and natural control. Chemical pest control in integrated farming is based on accurate sampling of pest population densities linked to economic thresholds. Using this approach, a consistent reduction of more than 90% in pesticide use can be achieved with the same economic return as conventional farming (van Lenteren 1997a). The generally lower yields from the integrated system are compensated by reductions in cost from the lower input of pesticides and fertilizers.

Biological control will be a key feature of future sustainable crop production (van Lenteren 1998). The landscape in which agriculture currently takes place is characterized by (i) low species diversity, (ii) plants with little architectural complexity, and (iii) plants and animals that are short-lived with high fecundity, relatively good dispersal but poor competitive ability (Bukovinszky 2004). Further, many agro-ecosystems are dominated by weeds, insects and pathogens that are highly adapted for rapid colonization and population increase. Plants with simple architectures have fewer associated species of insects (pests and beneficials) than diverse and architecturally complex plant communities (Landis & Marino 1997). As a consequence of these low-diversity plant and herbivore communities, agro-ecosystems often have strongly impoverished natural enemy communities when compared with natural ecosystems (Landis et al. 2000). However, the areas adjacent to crop fields are usually less disturbed and architecturally more complex, with a richer and more stable natural enemy fauna that can provide source populations of beneficial arthropods to promote pest management. But it should be realized that these extra-field communities may also act as a reservoir for pest species (Winkler 2005).

Sustainable pest management must, therefore, be based on an appreciation of how the structure of the agricultural landscape can influence the interactions between extra-field and within-field processes. An understanding of the interchange of organisms between areas of the landscape, and the influence of landscape structure on these interchanges, is critical for predicting and managing pest populations in agricultural fields (Lewis et al. 1997). However, as a starting point, it might be more efficient to first concentrate on changes within cropping systems that could increase natural pest control. One approach would be the use of multi- or poly-cropping systems with (i) two or more crop species, (ii) a crop with an undersowing of an economically unimportant plant, or (iii) a multi-crop consisting of a crop species and herbaceous field margins. Although there are many publications which state that the natural enemy fauna is richer in multi-crops, and that natural biological control is greater than in monocultures, there is very little experimental, quantitative data to support this view (see Vandermeer (1989) for a review). There is, for example, hardly any information on how natural enemies search for prey in multi-crop systems compared with searching behaviour in monocultures. However, one of the most frequently cited reasons for multi-cropping (which is applied on 60% of the world area used for food production) is protection from pests (Vandermeer 1989). Pest pressure is usually lower in multi-crops, though not always. The presence of associated plants in a multi-crop can reduce the damage to the main crop species in three ways, all involving a lower population growth rate of the pest. In the first situation, the associated plants are more attractive hosts for the pest than the crop (host-plant quality hypothesis). In the second response, the associated plants interfere directly with activities of the pest (disruptive crop hypothesis); and in the third, the associated plants change the environment to favour natural enemies of the pests (natural enemies hypothesis).

A series of studies have tried to identify the mechanisms by which the numbers of insect herbivores are reduced in multi-crop systems and concluded that in most cases the disruptive crop hypothesis is the best explanation for the data, but that the natural enemies hypothesis could often be supported (Risch et al. 1983). Some recent experimental studies indicate that all three hypotheses for pest reduction, the host-plant quality hypothesis (see Theunissen et al. 1995), the disruptive crop hypothesis (see Visser 1986; Finch & Kienegger 1997) and the natural enemies hypothesis (see Coll & Bottrell 1996) may be valid. Other reviews reveal similar results and data clearly show that plant diversity often results in higher natural enemy populations (e.g. Andov 1983).

An analysis of 51 recent studies of habitat manipulation to enhance conservation biological control (Gurr et al. 2000) showed that in the vast majority of cases there were significant benefits for natural enemies. However, a significantly beneficial effect on natural enemies did not always result in a greater reduction of pest populations or higher yields. Owing to the empirical approach that typifies many of these studies, the effects of agro-ecosystem diversification on the searching behaviour and success of arthropod natural enemies remains poorly understood. Studies in this area are a priority to inform the design and fine-tuning of farming schemes aimed at pest prevention.

4. BIOLOGICAL CONTROL AND GM CROPS

The estimated global area of commercially grown GM crops in 2004 was 81 Mha, equivalent to approximately 5% of the global area available for cultivating crops and 3.2 times the total land area of the UK (James 2004). Four major GM crops (soya bean, cotton, maize and oilseed rape) were grown by 8.25 million farmers in 17 countries in 2004, and approximately 90% of the farmers were resource poor from developing countries. GM herbicide tolerance in these four crops occupied 58.6 Mha (72% of the total GM area), while GM insect resistance in maize and cotton was grown on 15.6 Mha (19%). Stacked genes for herbicide tolerance and insect resistance in maize and cotton occupied 6.8 Mha (9%). GM crops containing genes of Bacillus thuringiensis (Bt) expressing truncated Cry proteins which are toxic to specific insect groups (e.g. Lepidoptera, Coleoptera) are currently the only commercial insect resistant GM plants and are grown on 22.4 Mha (28%). Bt Cry1-expressing maize and cotton are protected from attack by lepidopteran pests including corn borers (mainly O. nubilalis) in maize and the budworm-bollworm complex (H. virescens, Helicoverpa spp., Pectinophora gossypiella) in cotton (Shelton et al. 2002). In 2003, Bt maize for control of Diabroticaspp. (corn rootworms) was commercialized, expressing the coleopteran-specific Cry3Bb toxin.

An advantage of insect-resistant transgenic Bt plants is the reduced need for conventional insecticides, providing benefits for human health and the environment (Shelton et al. 2002). For example in US cotton, the average number of insecticide applications used against the budworm–bollworm complex decreased from 4.6 in 1992–1995 to 0.8 in 1999–2001, largely due to the introduction of Bt cotton (FAO 2004). In China, the introduction of Bt cotton has led to a 60–80% decrease in the use of foliar insecticides (Fitt et al. 2004).

Detailed analysis of potential effects of GM crops on the environment and the crop ecosystem is crucial prior to commercial release (Dale et al. 2002). Key concerns are potential ecological consequences on non-target organisms, including the natural enemies of pests. For Bt toxin, direct effects can be expected only if it is ingested and the parasitoid/predator is susceptible. This requires that the toxin is acquired through direct plant feeding (e.g. pollen), or that it is passed on in a biologically active form by the host/prey. Indirect, host/prey-quality mediated effects can be expected if susceptible herbivores ingest the toxin. Affected herbivores are often smaller, develop slower, behave differently and/or have an altered tissue composition compared to healthy individuals. Such changes may influence the quantity of prey and their nutritional quality for the next tropic level. Effects on natural enemies include death, sublethal effects (e.g. prolonged development, reduced weight), altered behaviour (e.g. reduced parasitization rate, changes in prey choice), or species may be unaffected.

Over the past 10 years, a number of laboratory studies have investigated possible deleterious effects of Bt plants on the mortality, longevity and development of predators and parasitoids. To date, there is no indication of direct effects of Bt transgenic plants on any of the species studied so far, either in direct plant-feeding assays or when the natural enemies had been provided with non-susceptible prey or hosts containing the Bt (Cry) proteins (Romeis et al. 2006). Adverse effects have only been reported in studies using herbivores as prey or hosts that ingested the Bt toxin and were susceptible to it (Romeis et al. 2006). Such indirect host-quality mediated effects are to be expected because sublethally intoxicated prey or hosts are generally of lower nutritional quality. However, such indirect effects are not unique to Bt toxins, and are known to occur with other insecticides.

On a larger scale, more than 50 studies have been conducted in both experimental and commercial fields to evaluate the impact of Bt crops on natural enemies. Experimental field studies have revealed only minor, transient or inconsistent effects of Bt crops when compared to a non-Bt control. Exceptions were observed with specialist natural enemies which were virtually absent in Bt fields due to the lack of target pests as prey (Riddick et al. 1998) or hosts (Pilcher et al. 2005). A few studies in Bt crops revealed consistent reductions in the abundance of different generalist predators that were also associated with the reduced availability of sensitive (target) lepidopteran prey (Daly & Buntin 2005; Naranjo 2005a; Whitehouse et al. 2005). Some studies have compared the function of natural enemies (parasitization and predation) in Bt and conventional crops; parasitization rates of naturally occurring or sentinel larvae of sensitive (target) lepidopteran species in Bt crops have been reported to be lower compared to control plots (Siegfried et al. 2001; Bourguet et al. 2002). This reduction in parasitism is not surprising given that host populations were significantly decreased by the Bt crop. Predation rates on sentinel lepidopteran eggs or larvae did not differ between Bt and untreated non-Bt fields (Musser & Shelton 2003; Sisterson et al. 2004; Naranjo 2005b), but were significantly reduced by the application of broad-spectrum insecticides (Musser & Shelton 2003). Studies in commercially managed cotton fields revealed much higher predation rates in Bt cotton compared to non-Bt fields, where more insecticides were applied (Head et al. 2005). A 6-year field study in Bt cotton on the abundance of 22 arthropod natural enemy taxa indicated that the exposure to the Bt toxin over multiple generations did not cause any chronic long-term effects (Naranjo 2005a). As yet, there is little evidence that secondary pest outbreaks in Bt crops have emerged as a problem requiring significant use of insecticides. This confirms that overall biological control is not negatively affected by the use of Bt plants (Fitt et al. 2004; Naranjo 2005b; Whitehouse et al. 2005). The analysis of all published peer-reviewed studies on Bt crops and natural enemies (Romeis et al. 2006) has indicated that parasitoid and predator abundance and level of biological control are similar in Bt and untreated non-Bt crops, whereas broad-spectrum insecticides generally lead to a drastic reduction in natural enemy populations as well as a loss of biological control function. It seems, therefore, that in those Bt crops that have been cultivated in commercial production, the Bt technology can contribute to natural enemy conservation and be a useful tool in IPM. Furthermore, it appears that Bt crops and biological control could be used in combination, which may be particularly valuable where the GM-conferred insect resistance is not effective against other key pests. As GM technology evolves, it will be important to evaluate GM traits conferring insect resistance on a case-by-case basis, to identify any adverse effects on the associated natural enemy fauna, and the compatibility of the GM crop with biological control.

5. Risks, Benefits and Regulation

Since pest problems in agriculture involve plants, plant-feeding organisms and their natural enemies, the regulation of biological control agents has usually been the responsibility of national plant quarantine services. For this reason, regulation over several decades focused on the need to ensure that introduced natural enemies would not become agricultural pests (Waage 1997). Concern about the risk which introduced biological control agents used against arthropod pests might pose to natural, non-agricultural ecosystems did not become a major issue until much more recently. In contrast, the potential for herbivorous arthropods to become pests when introduced for the biological control of weeds has been recognized for a long time, and pre-release screening procedures, such as host-specificity testing, have been in place for over 40 years (Harris & Zwoelfer 1968). Pre-release screening of entomophagous biological control agents lagged behind their weed counterparts until the warnings by Howarth (1983, 1991) and Lockwood (1993), which raised concerns about possible impacts on non-target species. Interestingly, an analysis of published and unpublished information found that only 1.5% of entomophagous biological control agents introduced before 1999 had undergone any pre-release host specificity testing (Lynch et al. 2001). Yet, despite the fact that non-target effects of entomophagous biological control agents were rarely considered prior to import and release, there are only a few known cases where any adverse effects can be attributed to the release of such organisms (van Lenteren et al. 2006a).

The increasing popularity of biological control as an alternative to pesticide-based programmes has resulted in more frequent import and release of exotic entomophagous organisms in many countries (van Lenteren 1997b). In the past, Europe has generally been a source rather than a recipient of invertebrate biological control agents, in comparison to other countries with greater experience in classical biological control such as Australia, Canada, New Zealand, South Africa and the USA (Greathead 1976). These countries had legislation and testing procedures in place relatively early to regulate imports and analyse the risks of exotic entomophagous biological control agents (Sheppard et al. 2003).

Increasing international trade in agricultural products together with accidental introductions of organisms related to tourism and global trade have become important sources for new imports of exotic pests as demonstrated by Bin & Bruni (1997) for Italy. Many of these introduced organisms are candidates for classical biological control, especially if they establish in conservation areas where they may threaten native species and communities. A significant event in the history of regulation of entomophagous biological control agents was the adoption of the Convention on Biological Diversity following the Rio Conference in 1992. Signatory countries have accepted the obligation to prevent the introduction and to control alien species that threaten indigenous ecosystems and habitats. A broad and ongoing discussion within the scientific community and in public fora was initiated by the publication of the Convention, and biodiversity issues made a sudden appearance on the agendas of policy makers. As a consequence, the public perception of threats to biodiversity has changed markedly in recent years. Biological control, which had previously been viewed as an environmentally friendly method of pest control was subjected to a more critical evaluation, the assumed safety was questioned, and the need for regulation of the import and release of non-native species has become a keenly debated issue.

(a) Harmonization of regulation of entomophagous biological control agents

Regulation of the introduction and release of entomophagous biological control agents differs between countries, and some have yet to establish guidelines and procedures. International laws and agreements require a harmonized and effective regulatory system between countries. In many cases, introductions of biological control agents are administered under regulations which were established for other purposes, such as plant quarantine, wildlife conservation and GM plants. The application of appropriate regulatory procedures is important in order to maintain public confidence in biological control and to facilitate introduction and use of non-native species.

A number of regulatory documents for the import and release of invertebrate biological control agents have been produced by international organizations; for example, the Food and Agricultural Organization of the United Nations (FAO 1996, 2005), the European and Mediterranean Plant Protection Organization (EPPO 1999, 2001, 2002) and the Organization for Economic Co-operation and Development (OECD 2004). More recently, guidance on procedures and methods for environmental risk assessment (ERA) of non-native invertebrate biological control agents has been proposed (van Lenteren et al. 2003; Babendreier et al. 2005; van Lenteren et al. 2006b; Bigler et al. 2006). The biological control industry was concerned when the OECD guidance document was published, as the information requirements were considered to be too stringent, and manufacturers feared that national authorities would establish their own regulatory systems. As a consequence, the International Biocontrol Manufacturers Association proposed that the International Organization for Biological Control of Noxious Animals and Plants/West Palaearctic Regional Section (IOBC/WPRS) should coordinate the harmonization of the various regulatory documents. This resulted in the establishment of a Commission of the IOBC/WPRS in 2003 with the aim of offering a platform to scientists, regulators and industry to develop coordinated guidelines for European countries. The recommendations of this Commission were recently published (Bigler et al. 2005), and can now be adopted and implemented by national authorities.

The regulatory guidelines propose a two-step procedure in which the first stage is to license the import of an organism for research under contained conditions. This enables researchers to identify the key biological characteristics of the organism, such as host range and climatic requirements, which determine the ability of species to establish in new environments. This phase of the regulatory process should provide sufficient information for judgements to be made on the safety and efficacy of the organism. In the second step, companies submit an application for a licence to release the organism, with a dossier of information compiled in line with the guidelines, and including an evaluation of the risks and benefits arising from the proposed release. The risk–benefit analysis should include both positive and negative effects to the economy (farmer, producer and public), human/animal health and the environment (ERMA NZ 2000).

In contrast to pesticides, biological control agents usually lack the potential to adversely affect human and animal health. There have been some rare cases of adverse effects on human health in allergic reactions to some organisms (for example, to wing scales of Lepidoptera) among personnel working in production facilities. Environmental risks and benefits of biological control extend to a comparative evaluation (with pesticides) of effects on water, soil, air, biodiversity and ecosystem functions. Invertebrate biological control agents do not pollute water, soil or air (Greathead 1995), and thus the assessment of potential environmental effects have focused on biodiversity issues. However, it is difficult to assign monetary values to the damage or loss of species or ecosystem functions (Simberloff & Stiling 1996; Thomas & Willis 1998), such that environmental effects have to be considered in a qualitative manner. As a generalization, a strong case can be made for the view that biological control contributes to a reduction in pesticide use with corresponding benefits to the environment. While it should not be assumed that introduced biological control agents pose no risk to native species and ecosystems, history suggests that there have been remarkably few problems, even through the many decades when regulation was virtually non-existent (Lynch et al. 2001; van Lenteren et al. 2006a). The recently published regulatory guidelines (Bigler et al. 2005) are intended to produce a careful analysis of the establishment potential of an imported species in a new environment and any adverse effects on non-target organisms and ecosystem functions that may then arise. Consequences of regulation for the use of biological control agents in sustainable agriculture

Stakeholders in the area of biological control include scientists, industrial producers, retailers, growers, consumers, environmentalists and government regulators. Different stakeholders may have overlapping or conflicting interests; for example, the biological control industry and national organizations with responsibility for protecting the natural environment. It can be a time-consuming exercise to find a consensus between these differing views, with the net effect that innovative products are kept off the market for a long time, or that some species are never released. The costs of delaying product introduction to the market or the failure to licence a species at all can be calculated in terms of the economic loss for producers, retailers and growers, whereas the benefits of effective regulation can be expressed as 'public goods', like environmental protection. When seen in these terms, it is clearly difficult to make direct comparisons between the costs and benefits of regulation.

Markets for biological control products are usually small and cannot support or justify high levels of expenditure on the evaluation of environmental risks and preparation of the dossiers that are required for registration. The procedure is not only costly but also time consuming; the dossier for the registration of the bacterium Pseudomonas chlororaphis used for seed treatment against seed-borne diseases of barley and wheat was submitted in January 1996 under European Union (EU) directive 91/414 and was finally approved in April 2004. The regulatory process can be very costly and therefore may prevent small and medium-sized enterprises from developing biological control products. Only larger companies have the necessary finance to conduct research, produce a dossier and enter into the iterative discussions that are part of the registration procedure. As a consequence of overly stringent regulations, small and medium-sized businesses may be discouraged or prevented from identifying and assessing new species as potential biological control agents. Unnecessary over-regulation thus inhibits the development and introduction of innovative products for sustainable agriculture, and also impacts on start-up and spin-off companies in the biological control sector, as venture capital becomes even more scarce in the light of reduced returns on investment.

The current dilemma is not whether to regulate biological control agents, but how to regulate without introducing negative trade-offs which prevent innovation and limit the introduction of novel biological control agents in sustainable crop protection systems. In essence, the regulatory framework, and therefore associated costs, should be set at the minimum level that does not compromise environmental safety and human health.

An efficient and low-cost regulatory system is likely to be characterized by a number of key features: (i) applicants and regulators should establish contact at an early stage in the regulatory process and exchange information openly thus preventing unnecessary data generation; (ii) harmonized regulation among countries (e.g. across Europe) will be cost-effective because companies will be able to submit the same (or slightly modified) risk assessment dossier in different countries; (iii) governments should support public–private partnership to gather information on risks and benefits of candidate agents; and (iv) when previous releases in other countries or ecoregions with similar climatic conditions have proved to be effective and environmentally safe, a light-touch process of notification rather than regulation should be applied.

There have been a number of recent developments that hold out the prospect of a harmonized regulatory system in Europe. Firstly, scientists and representatives of the biological control production industry collaborated in an EU-funded project (ERBIC, 'Evaluating Environmental Risks of Biological Control Introductions into Europe') to develop a structure for a general ERA, consisting of (i) characterization and identification of the biological control agent, and determination of (ii) risks to human health, (iii) environmental risks, and (iv) efficacy. The ERA was proposed to focus on five risk factors of natural enemies: host/prey range, establishment, dispersal, and direct and indirect effects of releases (van Lenteren et al. 2003). It was recognized that the design and application of an ERA depends crucially on the availability of appropriate methods that are both realistic and cost-effective for the industry to apply. In this respect, the recent discovery that the outdoor winter survival (and hence establishment potential) of a number of insect and mite agents used in greenhouses is strongly correlated with their low-temperature survival under laboratory conditions (Hatherly et al. 2005 and references therein) is an important contribution to a risk assessment protocol.

FUTURE PERSPECTIVES AND CONCLUSIONS

It may not be an easy task at the present time to convince farmers worldwide to adopt a systems approach to pest management and make greater use of biological control. Yet in an agricultural industry that is still dominated by pesticides, biological control has found its place in the form of augmentative releases, particularly for the management of pests that are difficult to control with insecticides. Each pest species has tens to hundreds of associated natural enemy species (parasitoids, predators and pathogens), and thus thousands of natural enemies still await discovery. During the past 40 years, the identification and pre-release evaluation of natural enemies—including ERAs—has improved greatly (van Lenteren & Manzaroli 1999; Bigler et

and complex. Theoretically, a minimum of two criteria are involved. The first is recognition. There may be preformed molecules in both host and parasite that can interact. Second, there must be metabolic changes in the host or pathogen or both that are triggered by the initial interaction step. Genetic mutations in either host or pathogen can change the specificity of molecular interactions or their ability to trigger metabolic change.

The following presents discussions on research directed toward possible mechanisms involved in recognition between host and pathogen and the metabolic changes that cause disease symptoms.

Molecular Determinants of Resistance and Susceptibility

It is widely held that some forms of resistance to fungal and bacterial pathogens are the result of a host plant's ability to synthesize chemicals that inhibit the growth and development of the pathogen. During infection by a pathogen, metabolic pathways in the plant are activated, leading to the detectable biosynthesis of the inhibitors. A major class of inhibitors, called phytoalexins, are primarily low-molecular-weight, secondary plant metabolites that possess wide-ranging activity against fungi and, to a lesser extent, bacteria. In the last two decades, more than 100 phytoalexins have been identified. The induction of the biosynthesis of phytoalexins, however, does not follow the specificity that most pathogens have for a specific cultivar. For example, phytoalexin synthesis can be induced by abiotic agents, such as wounding or other stress conditions, in both resistant and susceptible plants. Phytoalexins can be toxic to both virulent and avirulent pathogens. It appears that phytoalexin synthesis might be a general, nonspecific type of active resistance.

An alternate approach, the study of susceptibility, has revealed mechanisms that show a high degree of specificity. Many pathogens possess specific agents for virulence, such as toxins or enzymes, that determine the course of events in susceptible plants. In the last five years, six host-specific or host-selective toxins have been chemically characterized. These toxins affect only susceptible cultivars and are produced only by specific pathogens that can attack these same susceptible cultivars. One well-studied example is the toxin produced by the fungus Helminthosporium maydis, mentioned earlier. The H. maydis toxin disrupts energy generation in susceptible mitochondria that characterize the T cytoplasm of corn. Normal mitochondria are resistant and are unaffected by the toxin because they apparently lack a receptor site for it.

Genetic specificity also exists for resistance and susceptibility to plant viruses, but there is no information on how such genes act. With respect to plant viruses the term resistance is used rather loosely. Quite often only the appearance of disease symptoms is considered. Thus, a plant that supports virus replication but shows no symptoms is considered to be resistant because it superficially appears to be so. More correctly, that plant should be called tolerant.

Recent observations suggest that one type of resistance may involve the ability of viruses to spread from cell to cell in their hosts. The continuum of responses ranges from rapid and complete invasion of the whole plant by the virus to slow invasion to circumstances where the virus is unable to spread from an infected cell, even though it might replicate well there. Accumulating evidence indicates that viruses induce the synthesis of proteins that are necessary for the movement of viruses from cell to cell. The host, however, depending on its genotype, can in some way interfere with this protein. Although the process is poorly understood, it may be, in part, the basis of resistance of plants to viruses.

In a sense, viruses might be thought of as packages of genes; they are composed primarily of RNA or DNA, and they can replicate only in a favorable host cell environment. Studies of the interactions between vital RNA or DNA and genes in the host cell can lead not only to an understanding of how viruses function but also to the development of viruses as gene-carrying vectors for genetic engineering.

An improved understanding of the basic concepts controlling resistance and susceptibility will result from research based on interrelated approaches to the analysis of the genetics of these traits, the gene products, the structure of the genes, and the methods that will permit their transfer between organisms.

Genetics

Continued breeding studies and genetic analysis of resistance traits in host plants and virulence traits in pathogens provide the experimental systems needed to isolate and determine the properties of recognition molecules involved in susceptibility or active resistance, such as phytoalexin biosynthesis.

Single-gene changes that confer resistance against a pathogen exist and are used in crop breeding to develop improved cultivars. In other cases multiple genes appear to be involved in resistance, and complicate crop breeding. The growing collection of data on the genetics of host plants and particularly of pathogens needs to be strengthened. Such data are essential for identification of the genes that control the specificity of receptor molecules, which determine resistance or susceptibility to bacteria, fungi, or viruses. Genetic analysis of some important fungal pathogens, however, will be difficult because sexual reproduction does not occur, and the modes of genetic reassortment and inheritance are unknown.

Many genetic approaches are now being initiated. For example, single-pathogen genes responsible for disease reactions in two bacterial leaf-spot diseases, soybean blight and bacterial spot of tomato, are being isolated and cloned. These techniques have potential for wide application.

GENE PRODUCTS

The end product of most genes is a protein. There is little direct evidence for the role of any specific proteins in controlling interactions between a host and a pathogen. Many potential candidates, however, can be hypothesized. By analogy with animal systems, surface molecules, such as membrane glycoproteins, may interact with low-molecular-weight messenger molecules, such as small carbohydrates released from cell walls. Cell wall extracts from both hosts and pathogens have been shown to elicit some resistance responses. Both the hydrolytic enzymes that release carbohydrate fragments from cell walls and the enzymes involved in the biosynthesis of toxins or phytoalexins are gene products that may be selected for study.

Additional basic information is needed about the cellular interactions between host and pathogen during the onset of resistance reactions. For example, the precise mechanisms employed by phytoalexins to exert their effects on pathogens are unknown and need to be actively studied. Metabolic pathways for the biosynthesis of phytoalexins must be clarified, and other compounds associated with resistance need to be identified. The regulation and coordinated synthesis of the enzymes involved in these pathways must be detailed.

In addition, the phenomenon of acquired resistance in plants needs further study. Resistance can be localized or can occur throughout the plant. Systemic resistance, however, may be of more practical value. This phenomenon can appear after a host plant is inoculated with an avirulent strain of the bacterial, fungal, or vital pathogen. This exposure somehow induces resistance properties so that when the plant is subsequently challenged by one or more pathogenic strains, it will resist infection or exhibit only mild disease symptoms.

Acquired resistance is most actively being studied using Pseudomonas solanacearum, some strains of which cause wilt and stem rot in tobacco, ginger, potato, tomato, and banana. Other avirulent strains only induce resistance. The experimental approach is to find mutants of the avirulent strains that fail to induce the acquired resistance. A comparison of the gene libraries of the active with the inactive mutants could lead to the identification of the genes and gene products responsible for triggering the acquired resistance.

Gene Structure

Once the genes and gene products are identified, it is feasible to alter their activity by changing the structure of the gene itself. The tools of molecular genetics can be used to study both the structure and activity of pathogen genes for virulence and avirulence and host genes for resistance and susceptibility. Some progress has been made recently with bacterial pathogens, particularly in characterizing some virulence factors such as pectolytic enzymes. Much of the basic information on the molecular biology of fungal pathogens, however, is yet to be acquired.

The functions of proteins coded for by vital genomes must also be established to aid in the understanding of their possible roles in replication and pathogenesis.

It may now be possible to isolate genes for specific types of resistance, such as that characterized by so-called hypersensitive lesions. For example, certain plant species and cultivars respond to infection by a pathogen by rapidly undergoing cell necrosis at the site of infection. The hypersensitive lesion can effectively stop the spread of a virus or confine the bacterial or fungal pathogen. In the latter two cases, the pathogen then dies.

This response is controlled in most cases by a single, dominant gene in the host plant. One approach to study of the mechanism controlling development of the hypersensitive lesion would be to first isolate messenger RNA from infected plants—those induced to give a hypersensitive response and those with a suppressed hypersensitive response. The mRNA from the suppressed plants could be used to prepare complementary DNA. This complementary DNA should recognize and hybridize with all the mRNAs from induced plants, except for those involved in the hypersensitive response. In principle, the remaining free mRNAs could then be used to probe a gene library of the hypersensitive plant for the gene that they can hybridize with. This gene should be the one responsible for inducing the hypersensitive lesion.

Gene Transfer

The ultimate goal of research discussed in this section on genetics, gene products, and gene structure is the routine transfer and expression of genes for resistance in agriculturally useful plants. As noted in the earlier chapter on genetic engineering, some bacterial and vital pathogens may be developed as suitable carriers for the transfer of genes into host plants. Current and prospective vectors take advantage of naturally occurring, intimate associations between microorganisms and plants, both pathogenic and beneficial. An appreciable effort is needed to identify and obtain suitable vectors in addition to the one successful vector, the Agrobacterium Ti plasmid that can be used in some dicotyledonous plants.

The techniques necessary to manipulate vectors are available and will likely be refined and improved within the next few years. It is, unfortunately, the lack of knowledge of the basic biology of plants and of the function, transfer, and expression of genes that restricts progress in this area.

MOLECULAR BASIS OF CELLULAR DAMAGE IN SUSCEPTIBLE HOSTS

Although it may appear that research on cellular damage and disease symptoms is a subset of the research discussed previously on resistance and susceptibility, its intent is distinct, but of equal major importance. Research emphasis in this area will yield insights into the biochemical mechanisms that result in cellular damage, or disease, following successful pathogenic invasion. As yet there is no clear explanation of how major symptoms, such as the yellowing and loss of chlorophyll in chlorosis or the tumors, galls, and morphological changes caused by cellular growth distortion, are induced once a virulent pathogen becomes established in a tissue. It may be possible to ameliorate symptoms or prevent crop damage directly by treatment, if the biochemical details are known. The little-understood phenomenon of natural tolerance to disease is evidence that such treatment should be possible. Indeed, the study of natural tolerance may be a valuable guide for developing disease protection traits for crop improvement.

Easily observable disease symptoms, such as chlorosis, necrosis, and cellular growth distortions, can have a number of diverse causes. Therefore, it is not possible to make progress on such generalized disease symptoms without some indication of the kinds of pathogens involved. Some research approaches hold promise for establishing general scientific principles of host-pathogen interactions.

Mode of Action of Toxins

Research in the last decade on purification and structural characterization has led to an acceptance of the concept that toxins are the potent chemical agents of virulence in many important diseases caused by bacteria and fungi. Only a small number of toxins have been chemically identified. Even fewer have a postulated target or receptor site in the host cell, as was described earlier for Helminthosporium maydis. But even in these few cases, it is not known how interference with the target site leads to cell damage. Much additional research is needed on toxins—on their genetics, such as chromosomal versus plasmid inheritance; on their chemical structure; on the pathways of biosynthesis in pathogens; and on their biochemical effects and role in pathogenesis.

NUCLEIC ACID INTERACTIONS

It is clear that the mere replication of a virus or viroid within a plant does not determine whether that plant will be diseased. There are many examples of strains that produce a great deal of virus, but with very little damage to the host. On the other hand, some of the most serious plant diseases are caused by viruses that replicate very sparingly.

Viruses, with their small genomes, have too little genetic information to code for the variety of proteins necessary to account for the almost infinite number of symptom types. Thus, it seems likely that interactions between the nucleic acid of the pathogen and that of the host initiate the disease process. Viroids, which are RNA molecules that do not code for a protein product, can cause symptoms similar to those caused by viruses. This lends support to the supposition that viruses as well as viroids interact directly with the genome of the host plant.

Complete nucleic acid sequences are now available for several viroids; for satellite RNAs, which modify the symptoms of their carrier viruses; and for a few plant viruses. Complete complementary DNA clones have been made for some of these RNA agents and have been shown to be infectious. Because DNA is technically easier to modify than RNA, such DNA clones provide the opportunity to make site-specific modifications in the sequence of the nucleic acid by inserting or deleting short stretches of DNA. The effect of such changes on the agent's ability to infect and on the symptoms produced can then be determined.

Using current methods the nucleotide sequences responsible for the disease syndrome should be identified. Furthermore, these complementary DNA clones could also be used in hybridization studies to locate regions in the host genome where the host and the virus, satellite RNA, or viroid sequences interact. As knowledge of the fine structure of the host's genes increases, future studies should enable researchers to determine the specific genes and processes that are perturbed by the presence of the pathogen.

If the DNA clones themselves are not infectious, the cloned viral or viroid DNA can be transcribed back to RNA using any of several in vitro systems. Thus, site-specific modifications made in the DNA clone can be transcribed into the RNA to test the effect of such changes on infectivity and disease symptoms. In this manner, critical regions of the genome could be identified, which would aid in the understanding of their functions and possibly the functions of viral-coded proteins.

Bacterial Interactions

Bacteria that cause diseases in plants cause symptoms, at least in part, by the production of various metabolites. Relatively few of these substances have been identified. The metabolites include, but are not limited to, toxins, polysaccharides, pectic enzymes, and plant hormones. All the bacterial toxins identified to date appear to be general toxins affecting a wide spectrum of plants. Many of these plants are not considered to be host species for the bacterial pathogen producing the toxin.

Other bacterial metabolites appear to have specific effects on host plant species. Bacterial polysaccharides, which are associated with wilting of plants, can be released in amounts great enough to clog up transport between plant cells, and may act by disrupting plant cell membrane functions. Soft rots, for example, are the result of bacterial enzymes that degrade the cementing pectin layer between plant cells. The production of plant hormones by bacteria disrupts the endogenous hormone balance in the host plant and can be part of the mechanism leading to crown gall tumors and other abnormal growths.

The molecular and genetic bases of the synthesis of these pathogen metabolites and the basis of the symptoms they cause in the host plant are largely unknown. There is increasing knowledge, however, about the genetics of some of the bacterial virulence factors that contribute to the severity of a disease. For example, in crown gall, which is caused by Agrobacterium, both bacterial chromosomal and plasmid genes are known to be required for pathogenicity. The molecular genetics of crown gall is the most thoroughly studied of any plant disease.

In the genetic analyses of virulence in bacteria, two different approaches are currently being used. One is the introduction of transposons into virulent strains of bacteria to create avirulent mutants. The transposon is used as a probe to locate and isolate the turned-off virulence gene. DNA clones of virulence genes can be used for an analysis of gene products. The second approach is molecular and genetic analyses of known or suspected determinants of pathogenicity, such as cell surface components, hormones, toxins, and extracellular enzymes. Both approaches hold promise for the elucidation of the biochemical steps in pathogenesis. There is an essential need to have basic knowledge about the structure, function, and regulation of virulence factors in the pathogen to provide a basis for directed plant breeding and to design effective inducers of plant resistance.

RESEARCH STATUS

It is important to recognize that considerable expertise and training in molecular biology are necessary for many of the research approaches discussed in this section of the report. Progress is facilitated by individuals working together in groups. Interactions with researchers in other laboratories are important sources of intellectual stimulation as well as sources of technical expertise.

The tools of genetics and molecular biology offer some new methods for understanding the highly specific interactions between host and pathogen. Studies of the molecular aspects of plant pathology must receive high priority and emphasis within the ARS research programs on plant diseases.

Currently the ARS research centers are undertaking relatively little basic work in molecular plant pathology. The ARS does have a few strong research programs in virology and in viroids, but very little work at the molecular level is being conducted with bacteria or fungi. A single laboratory, at Beltsville, is studying plant mycoplasmas.

To strengthen programs in the molecular basis of plant diseases, research investigations should emphasize:

- The molecular bases of the factors that determine whether a host-pathogen pair will result in a resistant or a susceptible interaction.
- The basic concepts of the interaction between the host and the invading pathogen that result in disease. This should lead to novel methods of preventing damage from disease, including natural plant tolerance.
- The transfer of resistance traits to normally susceptible plants through the development and subsequent exploitation of vector systems that allow for gene transfer between plant species.

It is significant to note that very few laboratories in the world have undertaken studies to understand the molecular basis of plant diseases from gene identification to disease symptoms.

The reliance on chemical-based pesticides, the increasing occurrence of pesticide resistance, particularly in insect pests, and the potential of such agricultural chemicals for polluting the environment are of increasing concern. The search for ways to address these concerns has led to greater emphasis on biological control. Biological control methods involve the use of one organism to mitigate the undesirable effects of another. Two complementary approaches are (1) the identification, biological characterization, and genetic engineering of crop-enhancing microorganisms, especially those that can be applied to seeds or roots to promote improved growth or yield; and (2) the development of genetically engineered microorganisms to remove organic pesticide residues, such as herbicides.

Microbial Agents for Biological Control

Little commercialization of microorganisms for biological control has been done. Yet there is great potential for research and development in this area. Knowledge of the basic biology of viruses, bacteria, fungi, nematodes, insects, and weeds is essential for identifying and developing naturally occurring antagonists as biological control agents. This includes knowledge of the growth and metabolism of the organisms obtained from both laboratory and field studies. Fundamental knowledge of the biological basis of the control mechanism and the ecology of both organisms involved is needed to successfully manipulate them to full advantage.

Interactions among several disciplines will be necessary. Soil physicists, meteorologists, computer modeling experts, and analytical chemists, as well as biologists with expertise in areas such as ecology, microbiology, and genetics will be essential contributors. The potential opportunities to use pathogens and pests against terrestrial and aquatic weeds, and against pathogens of agronomic crops, forest trees, and ornamental plants are enormous. Biological control is not generally expected to substitute totally for chemical control, but will supplement or be integrated with it.

CONTROL OF PATHOGENS

The proven feasibility of using biological control in this area has spurred research. The bacterium Agrobacterium radiobacter is commercially used to prevent infection of susceptible plants by a related bacterium, A. tumefaciens, which causes tumorous galls to form on many plants. The fungus, Peniophora gigantea, is used to control another fungus, Hetero-basidion [=Fomes] annosum, which causes root rot of pine trees. In both these cases the control mechanism is not completely understood, but is believed to result from competition between the control microorganism and the disease-causing organism for specific binding sites on the host plant. Also, it may be that the control organism can elicit a resistance reaction in the host. In the case of A. radiobacter, an antibiotic, agrocin 84, produced by the A. radiobacter has been identified as the possible mechanism in that example of biological control.

Biological control is also illustrated by the epiphytic bacterium Pseudomonas syringae. Both pathogenic and nonpathogenic strains of this bacterium are known to synthesize ice-nucleating proteins. When these bacteria on the plant surface are killed by antibiotic treatments or displaced with mutants of P. syringae that have lost their ability to synthesize ice-nucleating proteins, the plant can tolerate chilling to -7°C without frost damage. Mutant strains have been produced both by mutagenesis using chemicals or ultraviolet irradiation and by removing the ice-nucleation protein gene using genetic engineering techniques.

Control of Insect Pests

Alternative strategies for the control of insect pests need to be developed to augment the chemical and biological approaches currently in use. Some success has been achieved with Bacillus thuringiensis, used commercially for the biological control of some insects. This bacterium, when ingested, is lethal to the caterpillar stage of many insects. The bacterium harbors a toxic crystalline structure that dissolves in the alkaline hind-gut of susceptible caterpillars, resulting in disruption of digestion and death.

The use and genetic manipulation of insect pathogenic bacteria and viruses constitute a promising but comparatively underdeveloped approach to insect control. The potential exists for genetically improving these organisms to increase their pathogenicity, either by enhancing existing pathogenic traits or by introducing desirable pathogenic characteristics.

Basic knowledge about potentially useful pathogens must be acquired. This includes identification of the pathogen and characterization of the insect host. The specificity between pathogen and host and the techniques for production and storage of candidate pathogens must also be studied. With this information the physiology, biochemistry, and genetics of the host-pathogen interaction can then be investigated. More specific areas of study include the molecular basis of processes such as recognition, virulence, and toxicity and the mechanisms regulating gene function during these interactions.

Progress in this line of research is apparent from the work of many laboratories worldwide. Candidate microorganisms identified by this research include baculoviruses and Bacillus thuringiensis. With recent developments in insect cell culture, some of the fundamental processes detailed here, in principle, can be directly probed in vitro with any of these microorganisms.

Control of Nematodes

Control of plant parasitic nematodes has been largely accomplished through the use of chemical nematocides, many of which have now been shown to be harmful to the environment and have been withdrawn from use. Biological control measures using resistant plant varieties and trap crops have been effective in some cases. A trap crop can stimulate the hatching of nematode eggs but does not support nematode growth, thus reducing nematode populations to harmless levels. More information is needed on the basic biology of nematodes to provide directed approaches to their control, using less toxic, target-specific substances. This might include the use of the hatching stimulants that are apparently produced by plants and trigger nemarode eggs to hatch. The growing nematodes then perish in the absence of a suitable host plant. Studies of nemarode pheromones and hormones could lead to methods for controlling reproduction or development.

PLANT HEALTH MICROORGANISMS

In recent years some information has been gathered on soil microorganisms, specifically, certain bacteria, that can improve plant vigor and contribute to increased yields. The mechanisms by which such bacteria exert these effects are essentially unknown, nor are their relationships to pathogens or other microorganisms in the environment well understood. Indeed, candidate organisms suited for particular crops remain to be identified and characterized. Such bacteria contribute a desirable and perhaps essential microflora for optimal plant growth. While a range of microflora is known to be essential for human health, virtually nothing on a comparable basis is known for plants.

Several mechanisms have been suggested that describe the effects of soil microorganisms on plant health. Beneficial microbes may produce antibiotics that inhibit the growth of pathogens, or they may be involved in the acquired resistance phenomenon. Recent evidence suggests that some plant growth-promoting bacteria produce siderophores, iron-chelating molecules, that restrict the availability of this essential element to pathogens and other members of the microflora.

Biological Degradation of Organic Pesticides

Timely and appropriate disposal of pesticide residues in water and soils is an important and attainable goal in routine agricultural production practices. The biological degradation of pesticides is theoretically feasible. For example, pseudomonads have been identified as being able to degrade the herbicide 2,4-dichlorophenoxyacetic acid (2,4-D) to innocuous compounds. Lack of knowledge of the chemistry, the fate of breakdown products, and the ecology of the organisms involved, however, is still a constraint to their use.

Both waste disposal of agricultural by-products and biomass reduction on an industrial scale are under intensive investigation. The processes are not commercially feasible as yet, however, because of low yields and organism management problems. These problems can be overcome using genetically engineered organisms, especially bacteria, that are currently more amenable to manipulation than other microorganisms.

Research Status

The ARS laboratories are among those contributing to progress in biological control of plant pathogens and insect pests. With increased emphasis, the ARS could be at the forefront of this research. The potential return for the ARS extends beyond the control of plant pathogens and insect pests; it would involve the development of general methodologies for gene transfer, cloning, and gene expression using microbial and insect systems. Basic research on the microflora of the rhizosphere is also an area that ARS can strengthen.

There is enormous potential for the identification, development, and application of microorganisms that can degrade pesticide residues and other toxic wastes. The ARS should expand its efforts—some of which are exemplary—in these areas. It is high-risk, long-range research and requires the multidisciplinary base that is already in place in some locations.

Specifically, the ARS should focus research toward:

- Exploring and identifying microbial agents that can control plant diseases and insect pests. Further, the agency should seek conventional genetic or recombinant-DNA technologies to make these agents more effective;
- Generating more knowledge of the basic biology of plant pathogenic nematodes to develop novel, nonpesticide means of control by perturbing reproduction and development; and
- Developing unique microorganisms that will promote plant health and others that can be used to detoxify or destroy organic pesticide pollutants.

Molecular Basis of Pesticide Action

Pesticides are major tools in the production of food and fiber and in the maintenance of high standards of veterinary, human, and plant health. Better pesticides are needed, relative to cost effectiveness, potency, selectivity, persistence, environmental impact, and safety for domestic animals, humans, and plants. Most of the early pesticides were discovered in industrial programs involving the synthesis and screening of thousands of synthetic chemicals for safe and effective molecules. The emphasis in current discovery efforts favors research on the natural chemicals produced by plants and microorganisms, such as occurred for the pyrethroids. Equally important are investigations into the molecular basis of pesticide action.

Advances in bioregulation research provide new vistas in seeking enzyme or receptor targets for pesticide action. Increasing fundamental knowledge of the function and regulation of communication systems within living organisms focuses attention on new targets and greatly facilitates the molecular design of optimal compounds for pest control. Greater diversity is needed in the targets for future pesticides, such as insecticides, herbicides, nematocides, and fungicides to avoid or minimize the impact of pesticide resistance and toxicity against non-target species. Susceptible and tolerant species often differ only in the sensitivity of their pesticide receptor site or their facility for detoxifying the pesticide.

A clear definition of the mechanisms involved will provide the background for the next generation of improved pesticides. New pesticides, in turn, provide unique probes to explore cellular entities such as enzymes, receptors, and membranes.

The molecular basis for metabolic activation and detoxification must be defined. Using this background knowledge genetic engineering can provide opportunities for modifying receptor sites and detoxification mechanisms for improved animal and crop safety.

RESEARCH STATUS

Research on the molecular basis of pesticide action is carried out in many laboratories within industry, universities, and the ARS. Industrial labs tend to focus on the modes of action of their proprietary compounds. Universities more often use pesticides as probes for physiological and pharmacological investigations. The ARS has placed considerable emphasis on the mechanism of pesticide action. The laboratory defining a new target often reaps the benefit of finding alternative agents working at the same site or in the same way.

Research on pesticide mode of action requires the creative teamwork of biochemists, chemists, and geneticists with adequate instrumentation and the appropriate environment to stimulate communication. This multi-disciplinary approach and the requisite personnel are now in place in several ARS laboratories. The ARS should increase its emphasis on the molecular basis of pesticide action, using the available expertise in microbial, plant, and insect physiology, biochemistry, and natural products chemistry. Success in this program will serve as the basis for improving animal health and for reducing crop losses during production and storage.

More specifically, the ARS should emphasize:

- Definition of the molecular basis for metabolic activation and detoxification of pesticides;
- Study of new targets for selective pesticide action;

- Identification of new natural chemicals important in regulating pest populations;
- Investigation of the basic molecular biology of vectors for gene transfer and elucidation of gene regulation in insects; and
- Continued research on both insect genetics and on natural products chemistry.

Insect Neurobiology and the Regulation of Development and Reproduction

The functional responsiveness of an insect is dependent on rapid chemical communications among its own cells and between the individual and other insects. Intercellular communication is mediated primarily by the nervous system, through substances such as neurotransmitters, neurohormones, and neuromodulators as well as by the endocrine system, through hormones. The endocrine system is closely coupled to the functioning of the nervous system. Communication between individuals is achieved through volatile chemicals called pheromones. Their production and action is mediated by the nervous system.

Insect Neurobiology

The function of the nervous system makes it a logical focus for investigations of alternative means of insect control that could potentially have considerable selectivity. Before investigations can be initiated, however, basic information about the function of the insect nervous system must be obtained, specifically, information about nervous processes involving chemical communication. This approach is the only potentially successful avenue to the solution of applied research problems. For this reason a research emphasis in fundamental insect neurobiology should be developed by the ARS.

Insect neurobiology is now experiencing a period of exponential growth. Despite the fact that the insect nervous system has been used for many years as a model for studying certain neurophysiological processes, basic research using modern techniques has only recently begun on the neurochemistry, neuroendocrinology, neurogenetics, and neuropharmacology of the insect nervous system.

For example, the number of identified insect peptides with neurohormonal activity is fewer than 20. Only 4 of these insect neurohormones have been purified and sequenced. These include the neurotransmitter/neuromodulator proctolin, the two adipokinetic hormones, and cardiac accelerator peptides. Proctolin is important in the stimulation of muscle contraction and is co-released with other neurotransmitters. The adipokinetic hormones mobilize lipid for its metabolism by muscle in insect flight, and the cardiac accelerator peptides control the heartbeat of the insect. It now appears that the structures of the prothoracicotropic hormones, the primary effectors of insect metamorphosis and the first hormone of neural origin described for any animal (1917), are finally being resolved. In addition, a new brain peptide that regulates the production of pheromones has been described and promises to introduce a renaissance in pheromone research.

Study of these and of yet-undiscovered hormones will aid in an understanding of the physiology of the insect, its growth and development. Such studies will also define the mechanisms by which the central nervous system integrates and regulates these processes. This understanding may allow scientists eventually to selectively manipulate the neuroendocrine system, and thus control insects by altering their ability to fly, curtailing metamorphosis, or disrupting sexual recognition. The study of neurohormones may not provide an immediate answer to insect control. The resulting knowledge, however, will provide scientists with the sound foundation necessary to propose and pursue new directed and applied research on the neural regulation of insect growth and development.

The top scientific priority for neurobiological research on insects is the elucidation of the mechanisms by which chemical communication directs and coordinates the growth, development, homeostasis, and reproduction of insects. The basic information still lacking includes the identification of neural regulators and an elucidation of their chemistry, synthesis, secretion, and metabolism.

Other opportunities for manipulation of insect pests include the neurohormones bursicon, diuretic hormone, and egg development neurotropic hormone. Bursicon causes the insect skeleton to harden. Inhibition of the secretion of this hormone would cause death. Manipulation of the diuretic hormone, which regulates water and salt balance, might also result in death, through ionic imbalance and dehydration. Secretion of egg development neurotropic hormone from the brain of the female mosquito is stimulated following a blood meal. The hormone indirectly causes the ovary to mature the eggs. Manipulation of this reproductive hormone would prevent the development of generations of mosquitoes.

These hormones are examples of the potential in this field. To realize this potential the hormones must be studied extensively at the chemical, molecular, and physiological level.

At this point a major research program encompassing the physiology, biochemistry, and molecular biology of these regulators can be initiated. Research should include the study of mechanisms of communication within the nervous system, between organs and organ systems, and between individuals of the same species. Studies of interorganismal communication should emphasize the neuroendocrine and neural bases of this process and relate this communication to behavioral patterns in nature.

Knowledge gained from such a fundamental research program in insect neurobiology could be used in conjunction with genetic engineering methodologies to investigate the basic molecular biology of vectors for gene transfer and to elucidate gene regulation in insects. These new technologies could also aid in mapping the insect genome, particularly the genes for regulatory peptides.

Peptides offer researchers an extremely important direct line of study; they probably are all products of single genes. An understanding of these gene products or polyprotein precursors and their posttranslational processing to a bioactive peptide is essential for the potential control of insects. (Posttranslational processing, which follows the translation of RNA, is proving to be a fundamental mechanism that determines the protein nature of the neurosecretion from a given cell.) A disruption of the synthesis or processing of neurohormones would be lethal.

The long-term goal of this research is modification of the normal function of the insect nervous system to affect viability. A research program on insect pathogens as vectors for gene transfer would clearly be important in achieving this objective.

ENDOCRINE REGULATION OF METAMORPHOSIS, DIAPAUSE, AND REPRODUCTION

The postembryonic development of the insect involves a series of dramatic physiological and biochemical transformations that culminates in its emergence from a pupa as an adult form with its own unique function. It is generally accepted that these transformations and their associated metabolic processes all are directly or indirectly under endocrine control, including production of hormones by neural tissue. The full extent of the role of the endocrine system is not completely known, mainly because of a lack of knowledge of the hormones involved, the molecular basis of the developmental and reproductive processes these hormones control, and their mechanisms of action. The progress made in this field in recent years has largely been at a descriptive level. Thus, basic research is needed to identify and chemically characterize insect hormones and to define at the molecular level both their physiological function and their mechanism of action.

Although some insect hormones, such as the sesquiterpenoid juvenile hormones and the ecdysteroids, have been intensively investigated, the extent of their involvement in regulating insect development and reproduction is only now being realized. They are known to exist as structural and functional families of molecules, each acting at a specific time during the life cycle of the insect. The multiple functions of these hormones provide multiple avenues for pursuing control of the insect. Substantial stantially more research is needed, both in the above-cited areas as well as on the mechanisms of their interaction at the level of the target gland and interendocrine feedback control. Research studies must be designed to show how these hormones regulate one another's synthesis and secretion to drive development and growth.

A virtually unknown family of insect regulators that control metamorphosis, diapause, and reproduction is the peptides. Only a few have thus far been identified, and as has proved to be the case with vertebrates, there are numerous peptide hormones involved in the control of embryogenesis, postembryonic development, reproduction, and homeostasis. These peptides need to be characterized, their physiological functions defined, and mechanisms of action elucidated.

The regulation of the synthesis, secretion, and metabolism of these insect hormones, whether peptide, steroid, or other chemical structure, is another relatively unexplored research area of considerable significance and potential application to the control of insects. The secretion of these hormones has consistently been shown to be precisely regulated, frequently in response to discrete environmental cues such as photo-period, temperature, and stress. The mechanisms by which these cues are transduced by the nervous system to elicit an endocrine response are important areas for basic research in insect neurobiology.

Knowledge of the regulation of insect development and reproduction is applicable to the manipulation of these systems for improved pest control. Some natural and synthetic chemicals, including insecticidal compounds, alter growth and development by inhibiting the biosynthesis or action of juvenile hormones or ecdysteroids and by governing the initiation and termination of diapause. Certain antibiotics and the highly insecticidal benzoylphenyl ureas interrupt chitin synthesis necessary for the formation of the insect cuticle or skeleton. Studies on insect genetics indicate the possibility of breeding sterile hybrids for use in pest control. Bacteria and other microorganisms producing insecticidal materials and the plant itself may also be modified by selection and genetic engineering to increase the impact of natural toxicants or feeding deterrents in host-insect pest interactions. Further development of insect cell cultures and vectors for gene transfer in insects may permit the introduction of deleterious effects into pest populations.

The benefit from research in insect neurobiology is not the potential control of insect pests alone. Although the insect is a relatively simple system structurally, it is functionally complex, much like that of vertebrates. An understanding of the insect endocrine system will lead to a further understanding of similar processes in all eukaryotic organisms.

Research Status

The ARS is recognized worldwide for developing the sterile insect release method of control and for investigations on insect genetics and ecdysteroids. The agency also is internationally recognized for natural products research, particularly pheromone chemistry, and the application to insect development and reproduction. This type of interdisciplinary research requires a coordinated team of entomologists, physiologists, biochemists, and chemists.

There are a number of ARS laboratories currently conducting excellent research on the physiological and chemical aspects of endocrine control of insect development and reproduction. By bolstering these existing programs with the appropriate additions of scientists skilled in protein chemistry, basic biochemistry, and the study of nuclear and membrane proteins as receptors, the ARS should be able to make substantial contributions to this research area.

Although the ARS is becoming increasingly more involved in fundamental insect neurobiological research, this program is not developing in a focused manner. While most of the research skills necessary for a major program in insect neurobiology—chemistry, neurophysiology, behavior, biophysics, and physiology—are already in place within the ARS, additional expertise in neurochemistry, peptide chemistry, and biochemistry (mechanistic aspects or chemical regulation), and immunology must be added. Generally, adequate instrumentation for this research exists within the ARS. Analytical facilities are needed, however, for peptide and neurotransmitter structural identification.

Of the few laboratories worldwide engaged in insect neurobiological research, a number are emerging as centers of excellence. The comparative paucity of such centers, however, means that relatively few neurobiological systems are currently being explored. Thus, the scientific opportunities in this field are enormous. Unfortunately, the lack of basic information has created a situation wherein the most important areas of research are high risk and will require considerable effort and resources. Such high-risk research is well suited for government-supported organizations like the ARS.

To date, a multidisciplinary program in insect neurobiology does not exist. The ARS has an opportunity to establish the first program of this kind. The success of such a program greatly depends upon the centralization of research at a single site, preferably near a university or another research institute that has a strong program in neurobiology.

ARS research should specifically focus on the following:

- Chemistry of the brain factors that control pheromone production and release, and their mechanisms of action;
- Neural regulation of the synthesis, processing, and secretion of cerebral pheromonotropic peptides;
- The endocrine basis of insect reproduction, in particular, identification of the cerebral neuropeptides involved and their target glands, and identification of the mechanisms regulating these glands;
- Mechanisms that regulate the synthesis of ecdysteroids and juvenile hormones, and the biosynthetic pathways of these two hormome families; and
- Interhormonal endocrine feedback; regulation of insect growth, development, and reproduction; and the roles and molecular mechanisms of the principal developmental hormones in regulating

9

CROP ROTATION AND MECHANICAL CONTROL

CROP ROTATION

Crop rotation, the successive cultivation of different crops in a specified order on the same fields, in contrast to a one-crop system or to haphazard crop successions.

Throughout human history, wherever food crops have been produced, some kind of rotation cropping appears to have been practiced. One system in central Africa employs a 36-year rotation; a single crop of finger millet is produced after a 35-year growth of woody shrubs and trees has been cut and burned. In the major food-producing regions of the world, various rotations of much shorter length are widely used. Some of them are designed for the highest immediate returns, without much regard for the continuing usefulness of the basic resources. Others are planned for high continuing returns with protected resources. The underlying principles for planning effective cropping systems began to emerge in the middle years of the 19th century.

Early experiments, such as those at the Rothamsted experimental station in England in the mid-19th century, pointed to the usefulness of selecting rotation crops from three classifications: cultivated row, close-growing grains, and sod-forming, or rest, crops. Such a classification provides a ratio basis for balancing crops in the interest of continuing soil protection and production economy. It is sufficiently flexible for adjusting crops to many situations, for making changes when needed, and for including go-between crops as cover and green manures.

Definition

Crop rotation is the systematic planting of different crops in a particular order over several years in the same growing space. This process helps maintain nutrients in the soil, reduce soil erosion, and prevents plant diseases and pests. There is no universally accepted rotation schedule as the types of plants in a particular farm or garden depend on the local soil, climate, and resources available. The length of rotation time between different plants will also vary depending on the needs of the gardener.

MAXIMUM YIELD EXPLAINS CROP ROTATION

When a single crop is grown in one field for many years in a row, the crop will cause the depletion of particular nutrients from the soil. This depletion of nutrients leads to poor plant health and lower crop yield.

With crop rotation, particular nutrients are replenished depending on the crops that are planted. For example, a simple rotation between a heavy nitrogen using plant (e.g., corn) and a nitrogen depositing plant (e.g., soybeans) can help maintain a healthy balance of nutrients in the soil. Crop rotation also prevents plant diseases and pests by exchanging crops that may be susceptible to a particular disease or pest with a crop that is not susceptible. For example, although corn is affected by corn rootworms, soybeans are not. The soybeans help suppress the pest so that the corn planted the following year will not be as adversely affected by it.

There is no limit to the number of crops in a rotation. Depending on the needs of the gardener, a larger rotation schedule may be implemented, and can include the rotation of animal feed crops like hay, clover, or oats.

Fields may also be used as animal pastures or allowed to lay fallow until the following year. For organic farms and gardens, crop rotation is essential as it helps limit the amount of fertilizers and pesticides normally required.

History

Agriculturalists have long recognized that suitable rotations – such as planting spring crops for livestock in place of grains for human consumption – make it possible to restore or to maintain a productive soil. Middle Eastern farmers practiced crop rotation in 6000 BC without understanding the chemistry, alternately planting legumes and cereals. In the Bible, chapter 25 of the Book of Leviticus instructs the Israelites to observe a "Sabbath of the Land". Every seventh year they would not till, prune or even control insects. The Roman writer, Cato the Elder (234 – 149 BC), recommended that farmers "save carefully goat, sheep, cattle, and all other dung".

Two-field system

Under a two-field rotation, half the land was planted in a year, while the other half lay fallow. Then, in the next year, the two fields were reversed. From the times of Charlemagne (died 814), farmers in Europe transitioned from a two-field crop rotation to a three-field crop rotation.

THREE-FIELD SYSTEM

From the end of the Middle Ages until the 20th century, Europe's farmers practised three-field rotation, dividing available lands into three parts. One section was planted in the autumn with rye or winter wheat, followed by spring oats or barley; the second section grew crops such as peas, lentils, or beans; and the third field was left fallow. The three fields were rotated in this manner so that every three years, a field would rest and be fallow. Under the two-field system, if one has a total of 600 acres (2.4 km) of fertile land, one would only plant 300 acres. Under the new three-field rotation system, one would plant (and therefore harvest) 400 acres. But the additional crops had a more significant effect than mere quantitative productivity. Since the spring crops were mostly legumes, they increased the overall nutrition of the people of Northern Europe.

Four-field rotation

Farmers in the region of Waasland (in present-day northern Belgium) pioneered a four-field rotation in the early 16th century, and the Britishagriculturist Charles Townshend (1674-1738) popularised this system in the 18th century. The sequence of four crops (wheat, turnips, barley and clover), included a fodder crop and a grazing crop, allowing livestock to be bred year-round. The four-field crop rotation became a key development in the British Agricultural Revolution. The rotation between arable and ley is sometimes called ley farming.

Modern developments

George Washington Carver (1860s - 1943) studied crop-rotation methods in the United States, teaching southern farmers to rotate soil-depleting crops like cotton with soil-enriching crops like peanuts and peas.

In the Green Revolution of the mid-20th century the traditional practice of crop rotation gave way in some parts of the world to the practice of supplementing the chemical inputs to the soil through topdressing with fertilizers, adding (for example) ammonium nitrate or urea and restoring soil pH with lime. Such practices aimed to increase yields, to prepare soil for specialist crops, and to reduce waste and inefficiency by simplifying planting and harvesting.

Crop choice

A preliminary assessment of crop interrelationships can be found in how each crop: (1) contributes to soil organic matter (SOM) content, (2) provides for pest management, (3) manages deficient or excess nutrients, and (4) how it contributes to or controls for soil erosion.

Crop choice is often related to the goal the farmer is looking to achieve with the rotation, which could be weed management, increasing available nitrogen in the soil, controlling for erosion, or increasing soil structure and biomass, to name a few. When discussing crop rotations, crops are classified in different ways depending on what quality is being assessed: by family, by nutrient needs/benefits, and/or by profitability (i.e. cash crop versus cover crop). For example, giving adequate attention to plant family is essential to mitigating pests and pathogens. However, many farmers have success managing rotations by planning sequencing and cover crops around desirable cash crops. The following is a simplified classification based on crop quality and purpose.

Row crops

Many crops which are critical for the market, like vegetables, are row crops (that is, grown in tight rows). While often the most profitable for farmers, these crops are more taxing on the soil Row crops typically have low biomass and shallow roots: this means the plant contributes low residue to the surrounding soil and has limited effects on structure. With much of the soil around the plant exposed to disruption by rainfall and traffic, fields with row crops experience faster break down of organic matter by microbes, leaving fewer nutrients for future plants.

In short, while these crops may be profitable for the farm, they are nutrient depleting. Crop rotation practices exist to strike a balance between short-term profitability and long-term productivity.

Legumes

A great advantage of crop rotation comes from the interrelationship of nitrogen fixing-crops with nitrogen demanding crops. Legumes, like alfalfa and clover, collect available nitrogen from the soil in nodules on their root structure. When the plant is harvested, the biomass of uncollected roots breaks down, making the stored nitrogen available to future crops. Legumes are also a valued green manure: a crop that collects nutrients and fixes them at soil depths accessible to future crops.

In addition, legumes have heavy tap roots that burrow deep into the ground, lifting soil for better tilth and absorption of water.

Grasses and cereals

Cereal and grasses are frequent cover crops because of the many advantages they supply to soil quality and structure. The dense and far-reaching root systems give ample structure to surrounding soil and provide significant biomass for soil organic matter.

Grasses and cereals are key in weed management as they compete with undesired plants for soil space and nutrients.

Green manure

Green manure is a crop that is mixed into the soil. Both nitrogen-fixing legumes and nutrient scavengers, like grasses, can be used as green manure. Green manure of legumes is an excellent source of nitrogen, especially for organic systems, however, legume biomass doesn't contribute to lasting soil organic matter like grasses do.

PLANNING A ROTATION

There are numerous factors that must be taken into consideration when planning a crop rotation. Planning an effective rotation requires weighing fixed and fluctuating production circumstances, including, but not limited to: market, farm size, labor supply, climate, soil type, growing practices, etc. Moreover, a crop rotation must consider in what condition one crop will leave the soil for the succeeding crop and how one crop can be seeded with another crop. For example, a nitrogen-fixing crop, like a legume, should always precede a nitrogen depleting one; similarly, a low residue crop (i.e. a crop with low biomass) should be offset with a high biomass cover crop, like a mixture of grasses and legumes.

There is no limit to the number of crops that can be used in a rotation, or the amount of time a rotation takes to complete. Decisions about rotations are made years prior, seasons prior, or even at the very last minute when an opportunity to increase profits or soil quality presents itself. In short, there is no singular formula for rotation, but many considerations to take into account.

Implementation

Crop rotation systems may be enriched by the influences of other practices such as the addition of livestock and manure, intercropping or multiple cropping, and organic management low in pesticides and synthetic fertilizers.

Incorporation of livestock

Introducing livestock makes the most efficient use of critical sod and cover crops; livestock (through manure) are able to distribute the nutrients in these crops throughout the soil rather than removing nutrients from the farm through the sale of hay. In systems where use of farm livestock would violate reservations growers or consumers may have about animal exploitation, efforts are made to surrogate this input through livestock in the soil, namely worms and microorganisms.

In Sub-Saharan Africa, as animal husbandry becomes less of a nomadic practice many herders have begun integrating crop production into their practice. This is known as mixed farming, or the practice of crop cultivation with the incorporation of raising cattle, sheep and/or goats by the same economic entity, is increasingly common. This interaction between the animal, the land and the crops are being done on a small scale all across this region. Crop residues provide animal feed, while the animals provide manure for replenishing crop nutrients and draft power. Both processes are extremely important in this region of the world as it is expensive and logistically unfeasible to transport in synthetic fertilizers and large-scale machinery. As an additional benefit, the cattle, sheep and/or goat provide milk and can act as a cash crop in the times of economic hardship.

Organic farming

Crop rotation is a required practice in order for a farm to receive organic certification in the United States. The "Crop Rotation Practice Standard" for the National Organic Programunder the U.S. Code of Federal Regulations, section §205.205, states that:

Farmers are required to implement a crop rotation that maintains or builds soil organic matter, works to control pests, manages and conserves nutrients, and protects against erosion. Producers of perennial crops that aren't rotated may utilize other practices, such as cover crops, to maintain soil health.

In addition to lowering the need for inputs by controlling for pests and weeds and increasing available nutrients, crop rotation helps organic growers increase the amount of biodiversity on their farms. Biodiversity is also a requirement of organic certification, however, there are no rules in place to regulate or reinforce this standard. Increasing the biodiversity of crops has beneficial effects on the surrounding ecosystem and can host a greater diversity of fauna, insects, and beneficial microorganism in the soil.< Some studies point to increased nutrient availability from crop rotation under organic systems compared to conventional practices as organic practices are less likely to inhibit of beneficial microbes in soil organic matter.

While multiple cropping and intercropping benefit from many of the same principals as crop rotation, they do not satisfy the requirement under the NOP.

Intercropping

Multiple cropping systems, such as intercropping or companion planting, offer more diversity and complexity within the same season or rotation, for example the three sisters. An example of companion planting is the inter-planting of corn with pole beans and vining squash or pumpkins. In this system, the beans provide nitrogen; the corn provides support for the beans and a "screen" against squash vine borer; the vining squash provides a weed suppressive canopy and discourages corn-hungry raccoons.

Double-cropping is common where two crops, typically of different species, are grown sequentially in the same growing season, or where one crop (e.g. vegetable) is grown continuously with a cover crop (e.g. wheat). This is advantageous for small farms, who often cannot afford to leave cover crops to replenish the soil for extended periods of time, as larger farms can. When multiple cropping is implemented on small farms, these systems can maximize benefits of crop rotation on available land resources.

Benefits

Agronomists describe the benefits to yield in rotated crops as "The Rotation Effect". There are many found benefits of rotation systems: however, there is no specific scientific basis for the sometimes 10-25% yield increase in a crop grown in rotation versus monoculture. The factors related to the increase are simply described as alleviation of the negative factors of monoculture cropping systems. Explanations due to improved nutrition; pest, pathogen, and weed stress reduction; and improved soil structure have been found in some cases to be correlated, but causation has not been determined for the majority of cropping systems.

Other benefits of rotation cropping systems include production cost advantages. Overall financial risks are more widely distributed over more diverse production of crops and/or livestock. Less reliance is placed on purchased inputs and over time crops can maintain production goals with fewer inputs. This in tandem with greater short and long term yields makes rotation a powerful tool for improving agricultural systems.

SOIL ORGANIC MATTER

The use of different species in rotation allows for increased soil organic matter (SOM), greater soil structure, and improvement of the chemical and biological soil environment for crops. With more SOM, water infiltration and retention improves, providing increased drought tolerance and decreased erosion.

Soil organic matter is a mix of decaying material from biomass with active microorganisms. Crop rotation, by nature, increases exposure to biomass from sod, green manure, and a various other plant debris. The reduced need for intensive tillage under crop rotation allows biomass aggregation to lead to greater nutrient retention and utilization, decreasing the need for added nutrients. With tillage, disruption and oxidation of soil creates a less conducive environment for diversity and proliferation of microorganisms in the soil. These microorganisms are what make nutrients available to plants. So, where "active" soil organic matter is a key to productive soil, soil with low microbial activity provides significantly fewer nutrients to plants; this is true even though the quantity of biomass left in the soil may be the same.

Soil microorganisms also decrease pathogen and pest activity through competition. In addition, plants produce root exudates and other chemicals which manipulate their soil environment as well as their weed environment. Thus rotation allows increased yields from nutrient availability but also alleviation of allelopathy and competitive weed environments.

Carbon sequestration

Studies have shown that crop rotations greatly increase soil organic carbon (SOC) content, the main constituent of soil organic matter. Carbon, along with hydrogen and oxygen, is a macronutrient for plants. Highly diverse rotations spanning long periods of time have shown to be even more effective in increasing SOC, while soil disturbances (e.g. from tillage) are responsible for exponential decline in SOC levels. In Brazil, conversion to no-till methods combined with intensive crop rotations has been shown an SOC sequestration rate of 0.41 tonnes per hectare per year.

In addition to enhancing crop productivity, sequestration of atmospheric carbon has great implications in reducing rates of climate change by removing carbon dioxide from the air.

Nitrogen fixing

Rotating crops adds nutrients to the soil. Legumes, plants of the family Fabaceae, for instance, have nodules on their roots which contain nitrogen-fixing bacteria called rhizobia. During a process called nodulation, the rhizobia bacteria use nutrients and water provided by the plant to convert atmospheric nitrogen into ammonia, which is then converted into an organic compound that the plant can use as its nitrogen source. It therefore makes good sense agriculturally to alternate them with cereals (family Poaceae) and other plants that require nitrates. How much nitrogen made available to the plants depends on factors such as the kind of legume, the effectiveness of rhizobia bacteria, soil conditions, and the availability of elements necessary for plant food.

Pathogen and pest control

Crop rotation is also used to control pests and diseases that can become established in the soil over time. The changing of crops in a sequence decreases the population level of pests by (1) interrupting pest life cycles and (2) interrupting pest habitat. Plants within the same taxonomic family tend to have similar pests and pathogens. By regularly changing crops and keeping the soil occupied by cover crops instead of lying fallow, pest cycles can be broken or limited, especially cycles that benefit from overwintering in residue. For example, root-knot nematode is a serious problem for some plants in warm climates and sandy soils, where it slowly builds up to high levels in the soil, and can severely damage plant productivity by cutting off circulation from the plant roots. Growing a crop that is not a host for root-knot nematode for one season greatly reduces the level of the nematode in the soil, thus making it possible to grow a susceptible crop the following season without needing soil fumigation.

This principle is of particular use in organic farming, where pest control must be achieved without synthetic pesticides.

Weed management

Integrating certain crops, especially cover crops, into crop rotations is of particular value to weed management. These crops crowd out weed through competition. In addition, the sod and compost from cover crops and green manure slows the growth of what weeds are still able to make it through the soil, giving the crops further competitive advantage. By removing slowing the growth and proliferation of weeds while cover crops are cultivated, farmers greatly reduce the presence of weeds for future crops, including shallow rooted and row crops, which are less resistant to weeds. Cover crops are, therefore, considered conservation crops because they protect otherwise fallow land from becoming overrun with weeds.

This system has advantages over other common practices for weeds management, such as tillage. Tillage is meant to inhibit growth of weeds by overturning the soil; however, this has a countering effect of exposing weed seeds that may have gotten buried and burying valuable crop seeds. Under crop rotation, the number of viable seeds in the soil is reduced through the reduction of the weed population.

Not only do weeds have a negative impact on quality, and yields of potatoes, but it can also slow down operations when the crop is mature, and ready to harvest. Weeds make farmers less efficient when harvesting, because weeds like bindweeds, and knotgrass, can become tangled in the equipment, resulting in a stop-and-go type of harvest.

PREVENTING SOIL EROSION

Crop rotation can significantly reduce the amount of soil lost from erosion by water. In areas that are highly susceptible to erosion, farm management practices such as zero and reduced tillage can be supplemented with specific crop rotation methods to reduce raindrop impact, sediment detachment, sediment transport, surface runoff, and soil loss.

Protection against soil loss is maximized with rotation methods that leave the greatest mass of crop stubble (plant residue left after harvest) on top of the soil. Stubble cover in contact with the soil minimizes erosion from water by reducing overland flow velocity, stream power, and thus the ability of the water to detach and transport sediment. Soil Erosion and Cill prevent the disruption and detachment of soil aggregates that cause macropores to block, infiltration to decline, and runoff to increase. This significantly improves the resilience of soils when subjected to periods of erosion and stress.

When a forage crop breaks down, binding products are formed that act like an adhesive on the soil, which makes particles stick together, and form aggregates. The formation of soil aggregates is important for erosion control, as they are better able to resist raindrop impact, and water erosion. Soil aggregates also reduce wind erosion, because they are larger particles, and are more resistant to abrasion through tillage practices.

The effect of crop rotation on erosion control varies by climate. In regions under relatively consistent climate conditions, where annual rainfall and temperature levels are assumed, rigid crop rotations can produce sufficient plant growth and soil cover. In regions where climate conditions are less predictable, and unexpected periods of rain and drought may occur, a more flexible approach for soil cover by crop rotation is necessary. An opportunity cropping system promotes adequate soil cover under these erratic climate conditions.In an opportunity cropping system, crops are grown when soil water is adequate and there is a reliable sowing window. This form of cropping system is likely to produce better soil cover than a rigid crop rotation because crops are only sown under optimal conditions, whereas rigid systems are not necessarily sown in the best conditions available.

Crop rotations also affect the timing and length of when a field is subject to fallow. This is very important because depending on a particular region's climate, a field could be the most vulnerable to erosion when it is under fallow. Efficient fallow management is an essential part of reducing erosion in a crop rotation system. Zero tillage is a fundamental management practice that promotes crop stubble retention under longer unplanned fallows when crops cannot be planted. Such management practices that succeed in retaining suitable soil cover in areas under fallow will ultimately reduce soil loss. In a recent study that lasted a decade, it was found that a common winter cover crop after potato harvest such as fall rye can reduce soil run-off by as much as 43%, and this is typically the most nutritional soil.

Biodiversity

Increasing the biodiversity of crops has beneficial effects on the surrounding ecosystem and can host a greater diversity of fauna, insects, and beneficial microorganisms in the soil.Some studies point to increased nutrient availability from crop rotation under organic systems compared to conventional practices as organic practices are less likely to inhibit of beneficial microbes in soil organic matter, such as arbuscular mycorrhizae, which increase nutrient uptake in plants. Increasing biodiversity also increases the resilience of agro-ecological systems.

Farm productivity

Crop rotation contributes to increased yields through improved soil nutrition. By requiring planting and harvesting of different crops at different times, more land can be farmed with the same amount of machinery and labour.

Risk management

Different crops in the rotation can reduce the risks of adverse weather for the individual farmer.

Challenges

While crop rotation requires a great deal of planning, crop choice must respond to a number of fixed conditions (soil type, topography, climate, and irrigation) in addition to conditions that may change dramatically from year to the next (weather, market, labor supply). In this way, it is unwise to plan crops years in advance. Improper implementation of a crop rotation plan may lead to imbalances in the soil nutrient composition or a buildup of pathogens affecting a critical crop. The consequences of faulty rotation may take years to become apparent even to experienced soil scientists and can take just as long to correct.

Many challenges exist within the practices associated with crop rotation. For example, green manure from legumes can lead to an invasion of snails or slugs and the decay from green manure can occasionally suppress the growth of other crops.

Mechanical Control

Mechanical control is the management of pests by physical means such as the use of a barrier (e.g., screens or row covers), trapping, weeding or removal of the pest by hand. It may also involve changing the physical conditions in a given area, for example, changing the temperature to make an area unfavorable for pests.

Mechanical control means controlling pests and diseases with the help of mechanical measures.

1. Keep insects away from the plants. Many flying insects (aphids, butterflies, flies, capsids, thrips, etc.) can be kept away from the crop with the help of insect netting. By covering the soil or substrate with polythene, cloth or with special collars around the stem base it is possible to protect plants against larvae that eat roots or sub-soil stems. These measures also stop the development of larvae and pupae that need soil for completion of their life cycle, and keep them from further spreading.
2. Trap insects. With the help of sticky traps, insect-o-cutors, trap plants, pheromone traps etc. it is possible to trap winged insects.
3. Use temperature treatments to kill harmful organisms. There are different methods:

Hot waterSeeds, bulbs, tubers and cuttings can be immersed in hot water to kill potential pests such as insects, mites, nematodes, fungi and bacteria.

Hot airA hot air treatment can also kill harmful organisms in plants, bulbs and seeds.

SolarisationIf the soil is covered with transparent polythene for several weeks in summer, temperatures can rise so high through solar radiation that pest organisms are killed.

SteamingTreatments with steam will disinfect soil, substrate, crates, etc.

4. Use flooding as a technique. If a piece of ground can be flooded for a sufficient period of time, most of the harmful organisms will die due to lack of oxygen.
5. Remove infested plant materials. Place them in a bag, and destroy.

Sometimes the most efficient way to kill insects is to stomp on them, literally or figuratively! Physical or mechanical control methods can be as simple as hand-picking the bagworms from a juniper bush, cutting tent caterpillars out of a shade tree, or using a fly swatter and window screens to keep your home free of flying insects. At the other end of the technology spectrum are the electronic bug killers. These high-tech fly swatters produce an ultraviolet glow that attracts flying insects to an untimely death on an electrified grid. Although bug zappers probably kill more beneficial insects than pests, their owners seem to sleep better at night with the reassuring sound of bugs sizzling on the grill. From its inception in the late

1970's, the market for these devices has grown into an industry with annual sales approaching $100 million.

Simple entrapment devices work quite well to control some types of insects. Fly paper and sticky boards, for example, are often used in greenhouses to control whiteflies or leafhoppers. Fruit and shade trees can be protected from various pests (e.g., plum curculio, gypsy moth, and codling moth) by tying a band of folded burlap around the trunk with its open side facing down. As insects climb up the trunk, they are waylaid in the folds of burlap which can be treated with insecticide or inspected daily to collect the pests. Ditches or moats with steep vertical walls are occasionally used as barriers to keep crawling insects (e.g., chinch bugs or whitefringed beetles) from migrating out of one field and into another. Pitfall traps are dug at 3-5 meter intervals in the ditch and filled with kerosene or creosote to kill the pests.

In the 1930's and 1940's it was common practice for a farmer to hitch a bale of chicken wire behind a team of horses (or a tractor) and drive it through his fields to stir up dust. When the dust settled, it gave the insect pests an abrasive coating that gradually rubbed away their cuticular waxes and caused them to die from dehydration. Although this is certainly not the most reliable method of pest control, it does explain why crops planted near the edge of a well-travelled dirt road often have less insect injury than those on the opposite side of the same field.

Temperature extremes can be used to kill insects or prevent their injury. Cold storage of agricultural products prolongs their shelf life and retards the development of insect pests. Heat treatments are sometimes used in place of fumigation to kill insect larvae in certain types of produce. Mangoes, for example, are submersed in hot water baths (115°F for 68 minutes) to kill the eggs and larvae of fruit flies (Tephritidae) prior to export. In the 1950's and 1960's enterprising farmers in the Midwest built tractor-driven flame throwers that used liquid propane (LP gas) to ignite and burn off the stubble from corn fields in the hope that a variety of soil dwelling pests would be killed by the heat.

Unlike most of the other types of pest control discussed in this chapter, physical and mechanical tactics often become less practical (or less economical) when they are used on a large scale. Banding trees with burlap as protection against codling moth, for example, might be a reasonable approach for homeowners with five apple trees, but it would be far too time-consuming for commercial growers with 500 or 5000 trees.

THREE GOALS OF MECHANICAL PEST CONTROLS

Goal 1: Exclusion

The idea is pretty straight-forward: keep pests away! Exclusion can be achieved by several means, one of which is row covers. Row covers are not only useful for frost protection and weed control, but can provide a barrier from insect pests.

Another exclusion method is use of a mesh screen around plants or garden beds. If you grow in a greenhouse, you already have an exclusion method in place: your greenhouse walls. However: don't sink into a false sense of security- insect pests are sneaky! Your greenhouse will keep many insects out, but a few will find their way in, and as per the cruel humor of nature, the two insects that find their way in will be a fertile male and female with ambitions of setting a Guinness world record for most eggs laid. The bottom line is: don't rely on your greenhouse walls to keep out pests altogether.

Goal 2: Destruction

While working in a greenhouse a few years back, I was instructed in a particularly... gooey destruction method against slugs involving scissors and many a disgusted grimace. The "cut them in half" method was unpleasant but certainly effective, and a perfect example of destruction. Another common destruction method is the use ofdiatomaceous earth to kill insect pests. Destruction methods are unique to the pest and the situation (and whether you mind getting slug guts on your hands). Use your imagination! And if you can't figure anything out, recruit some young boys to kill some insects for you.

GOAL 3: REMOVAL

Removal methods include handpicking (which is usually followed by destruction), use of traps,water pressure sprays, and insect vacuums. Light traps, color traps, and sticky traps are all useful not only for collecting the offensive insects, but to monitoring pest population and the species that are present. Sticky traps are cheap and easy to use.

Methods

Handpicking

The use of human hands to remove harmful insects or other toxic material is often the most common action by gardeners. It is also classified as the most direct and the quickest way to remove clearly visible pests. However, it also has equal disadvantages as it must be performed before damage to the plant has been done and before the key development of insects.

Mechanical traps

Mechanical traps or physical attractants are used in three main ways: to efficiently trap insects, to kill them or to estimate how much many insects there are in the total landmassusing sampling method. However, some traps are expensive to produce and can end up benefiting insects rather than harming them.

Differences from integrated pest control

Integrated pest control refers to the use of any means to control pests once they reach unacceptable levels. Mechanical pest control is but a minor part of integrated pest control. It means only the use of physical means to control pests.

10

BIOLOGICAL INSECTS CONTROL

Biological control, biocontrol, or biological pest control is a method of suppressing or controlling the population of undesirable insects, other animals, or plants by the introduction, encouragement, or artificial increase of their natural enemies to economically non–important levels. It is an important component of integrated pest management (IPM) programs (Weeden et al. 2007).

The biological control of pests and weeds relies on predation, parasitism, herbivory, or other natural mechanisms. Therefore, it is the active manipulation of natural phenomena in serving human purpose, working harmoniously with nature. A successful story of biological control of pests refer to the human beings' capability to depict natural processes for their use and can be the most harmless, non–polluting, and self–perpetuating control method.

In biological control, the reduction of pest populations is achieved by actively using natural enemies.

Natural enemies of the pests, also known as biological control agents, include predatory and parasitoidal insects, predatory vertebrates, nematode parasites, protozoan parasites, and fungal, bacterial, as well as viral pathogens (Metcalf et al. 1973). Biological control agents of plant diseases are most often referred to as antagonists. Biological control agents of weeds include herbivores and plant pathogens. Predators, such as lady beetles and lacewings, are mainly free–living species that consume a large number of prey during their lifetime. Parasitoids are species whose immature stage develops on or within a single insect host, ultimately killing the host. Most have a very narrow host range. Many species of wasps and some flies are parasitoids. Pathogens are disease–causing organisms including bacteria, fungi, and viruses. They kill or debilitate their host and are relatively specific to certain pest or weed groups.

HISTORY

The term "biological control" was first used by Harry Scott Smith at the 1919 meeting of the Pacific Slope Branch of the American Association of Economic Entomologists, in Riverside, California. It was brought into more widespread use by the entomologist Paul H. DeBach (1914–1993) who worked on citrus crop pests throughout his life. However, the practice has previously been used for centuries. The first report of the use of an insect species to control an insect pest comes from "Nan Fang Cao Mu Zhuang" (WS[1]eI*f*(gÀr Plants of the Southern Regions) (ca. 304 AD), attributed to Western Jin dynasty botanist Ji Han (G]+T, 263–307), in which it is mentioned that "Jiaozhi people sell ants and their nests attached to twigs looking like thin cotton envelopes, the reddish-yellow ant being larger than normal. Without such ants, southern citrus fruits will be severely insect-damaged". The ants used are known as huang gan (huang = yellow, gan = citrus) ants (Oecophylla smaragdina). The practice was later reported by Ling Biao Lu Yi (late Tang Dynasty or Early Five Dynasties), in Ji Le Pian by Zhuang Jisu (Southern Song Dynasty), in the Book of Tree Planting by Yu Zhen Mu (Ming Dynasty), in the book Guangdong Xing Yu (17th century), Lingnan by Wu Zhen Fang (Qing Dynasty), in Nanyue Miscellanies by Li Diao Yuan, and others.

Biological control techniques as we know them today started to emerge in the 1870s. During this decade, in the USA, the Missouri State Entomologist C. V. Riley and the Illinois State Entomologist W. LeBaron began within-state redistribution of parasitoids to control crop pests. The first international shipment of an insect as biological control agent was made by Charles V. Riley in 1873, shipping to France the predatory mites Tyroglyphus phylloxera to help fight the grapevine phylloxera (Daktulosphaira vitifoliae) that was destroying grapevines in France. The United States Department of Agriculture (USDA) initiated research in classical biological control following the establishment of the Division of Entomology in 1881, with C. V. Riley as Chief. The first importation of a parasitoidal wasp into the United States was that of the braconid Cotesia glomerata in 1883–1884, imported from Europe to control the invasive cabbage white butterfly, Pieris rapae. In 1888–1889 the vedalia beetle, Rodolia cardinalis, a lady beetle, was introduced from Australia to California to control the cottony cushion scale, Icerya purchasi. This had become a major problem for the newly developed citrus industry in California, but by the end of 1889 the cottony cushion scale population had already declined. This great success led to further introductions of beneficial insects into the USA.

In 1905 the USDA initiated its first large-scale biological control program, sending entomologists to Europe and Japan to look for natural enemies of the gypsy moth, Lymantria dispar dispar, and brown-tail moth, Euproctis chrysorrhoea, invasive pests of trees and shrubs. As a result, nine parasitoids (solitary wasps) of gypsy moth, seven of brown-tail moth, and two predators of both moths became established in the USA. Although the gypsy moth was not fully controlled by these natural enemies, the frequency, duration, and severity of its outbreaks were reduced and the program was regarded as successful. This program also led to the development of many concepts, principles, and procedures for the implementation of biological control programs.

Prickly pear cacti were introduced into Queensland, Australia as ornamental plants, starting in 1788. They quickly spread to cover over 25 million hectares of Australia by 1920, increasing by 1 million hectares per year. Digging, burning and crushing all proved ineffective. By 1914, two control agents were introduced to help control the spread of the plant, the cactus moth Cactoblastis cactorum, and the scale insect Dactylopius. By 1917, most areas of prickly pear had been destroyed.

The first reported case of a classical biological control attempt in Canada involves the parasitoidal wasp Trichogramma minutum. Individuals were caught in New York State and released in Ontario gardens in 1882 by William Saunders, trained chemist and first Director of the Dominion Experimental Farms, for controlling the invasive currantworm Nematus ribesii. Between 1884 and 1908, the first Dominion Entomologist, James Fletcher, continued introductions of other parasitoids and pathogens for the control of pests in Canada.

Strategies of biological control methods

There are three basic types of biological control strategies; conservation biocontrol, classical biological control, and augmentative biological control (biopesticides).

Conservation biocontrol

The conservation of existing natural enemies is probably the most important and readily available biological control practice available to homeowners and gardeners. Natural enemies occur in all areas, from the backyard garden to the commercial field. They are adapted to the local environment and to the target pest, and their conservation is generally simple and cost–effective. For example, snakesconsume a lot or rodent and insect pests that can be damaging to agricultural crops or spread disease. Dragonflies are important consumers of mosquitoes.

Eggs, larvae, and pupae of Helicoverpa moths, the main insect pests of cotton, are all attacked by many beneficial insects and research can be conducted in identifying critical habitats, resources needed to maintain them, and ways of encouraging their activity (Lawrence 2005). Lacewings, lady beetles, hover fly larvae, and parasitized aphid mummies are almost always present in aphid colonies. Fungus–infected adult flies are often common following periods of high humidity. These naturally occurring biological controls are often susceptible to the same pesticides used to target their hosts. Preventing the accidental eradication of natural enemies is termed simple conservation.

Classical biological control

Classical biological control is the introduction of exotic natural enemies to a new locale where they did not originate or do not occur naturally. This is usually done by government authorities.

In many instances, the complex of natural enemies associated with an insect pest may be inadequate. This is especially evident when an insect pest is accidentally introduced into a new geographic area without its associated natural enemies. These introduced pests are referred to as exotic pests and comprise about 40 percent of the insect pests in the United States. Examples of introduced vegetable pests include the European corn borer, one of the most destructive insects in North America.

To obtain the needed natural enemies, scientists have utilized classical biological control. This is the practice of importing, and releasing for establishment, natural enemies to control an introduced (exotic) pest, although it is also practiced against native insect pests. The first step in the process is to determine the origin of the introduced pest and then collect appropriate natural enemies associated with the pest or closely related species. The natural enemy is then passed through a rigorous quarantine process, to ensure that no unwanted organisms (such as hyperparasitoids or parasites of the parasite) are introduced, then they are mass produced, and released. Follow–up studies are conducted to determine if the natural enemy becomes successfully established at the site of release, and to assess the long–term benefit of its presence.

There are many examples of successful classical biological control programs. One of the earliest successes was with the cottony cushion scale (Icerya purchasi), a pest that was devastating the California citrus industry in the late 1800s. A predatory insect, the Australian lady beetle or vedalia beetle (Rodolia cardinalis), and a parasitoid fly were introduced from Australia. Within a few years, the cottony cushion scale was completely controlled by these introduced natural enemies (Metcalf et al. 1973). Damage from the alfalfa weevil, a serious introduced pest of forage, was substantially reduced by the introduction of several natural enemies like imported ichnemonid parasitoid Bathyplectes curculionis. About twenty years after their introduction, the alfalfa area treated for alfalfa weevil in the northeastern United States was reduced by 75 percent (Metcalf et al. 1973). A small wasp, Trichogramma ostriniae, introduced from China to help control the European corn borer (Pyrausta nubilalis), is a recent example of a long history of classical biological control efforts for this major pest. Many classical biological control programs for insect pests and weeds are under way across the United States and Canada.

Classical biological control is long lasting and inexpensive. Other than the initial costs of collection, importation, and rearing, little expense is incurred. When a natural enemy is successfully established it rarely requires additional input and it continues to kill the pest with no direct help from humans and at no cost. Unfortunately, classical biological control does not always work. It is usually most effective against exotic pests and less so against native insect pests. The reasons for failure are often not known, but may include the release of too few individuals, poor adaptation of the natural enemy to environmental conditions at the release location, and lack of synchrony between the life cycle of the natural enemy and host pest.

Augmentative biological control

This third strategy of biological control method involves the supplemental release of natural enemies. Relatively few natural enemies may be released at a critical time of the season (inoculative release) or literally millions may be released (inundative release). Additionally, the cropping system may be modified to favor or augment the natural enemies. This latter practice is frequently referred to as habitat manipulation.

An example of inoculative release occurs in greenhouse production of several crops. Periodic releases of the parasitoid, Encarsia formosa, are used to control greenhouse whitefly, and the predaceous mite, Phytoseilus persimilis, is used for control of the two–spotted spider mite. The wasp Encarsia formosa lays its eggs in young whitefly "scales," turning them black as the parasite larvae pupates. Ideally it is introduced as soon as possible after the first adult whitefly are seen. It is most effective when dealing with low level infestations, giving protection over a long period of time. The predatory mite, Phytoseilus persimilis, is slightly larger than its prey and has an orange body. It develops from egg to adult twice as fast as the red spider mite and once established quickly overcomes infestation.

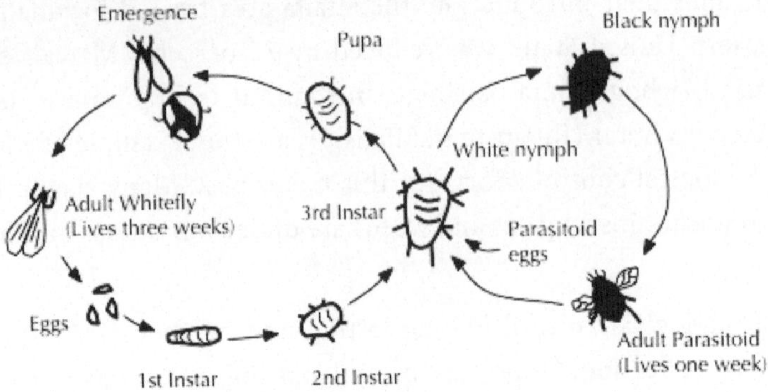

Fig 10.1

Diagram illustrating the life cycles of Greenhouse whitefly and its parasitoid wasp Encarsia formosa

Lady beetles, lacewings, or parasitoids such as Trichogramma are frequently released in large numbers (inundative release) and are often known as biopesticides. Recommended release rates for Trichogramma in vegetable or field crops range from 5,000 to 200,000 per acre per week depending on level of pest infestation. Similarly, entomoparasitic nematodes are released at rates of millions and even billions per acre for control of certain soil-dwelling insect pests. Entomopathogenic fungus Metarhizium anisopliae var. acridum, which is specific to species of short–horned grasshoppers (Acridoidea and Pyrgomorphoidea) widely distributed in Africa, has been developed as inundative biological control agent (LUBILOSA 2004).

Habitat or environmental manipulation is another form of augmentation. This tactic involves altering the cropping system to augment or enhance the effectiveness of a natural enemy. Many adult parasitoids and predators benefit from sources of nectar and the protection provided by refuges such as hedgerows, cover crops, and weedy borders. Mixed plantings and the provision of flowering borders can increase the diversity of habitats and provide shelter and alternative food sources. They are easily incorporated into home gardens and even small-scale commercial plantings, but are more difficult to accommodate in large–scale crop production. There may also be some conflict with pest control for the large producer because of the difficulty of targeting the pest species and the use of refuges by the pest insects as well as natural enemies.

Examples of habitat manipulation include growing flowering plants (pollen and nectar sources) near crops to attract and maintain populations of natural enemies. For example, hover fly adults can be attracted to umbelliferous plants in bloom.

Biological control experts in California have demonstrated that planting prune trees in grape vineyards provides an improved overwintering habitat or refuge for a key grape pest parasitoid. The prune trees harbor an alternate host for the parasitoid, which could previously overwinter only at great distances from most vineyards. Caution should be used with this tactic because some plants attractive to natural enemies may also be hosts for certain plant diseases, especially plant viruses that could be vectored by insect pests to the crop. Although the tactic appears to hold much promise, only a few examples have been adequately researched and developed.

DIFFERENT TYPES OF BIOLOGICAL CONTROL AGENTS

Predators

Ladybugs, and in particular their larvae which are active between May and July in the northern hemisphere, are voracious predators of aphids such as greenfly and blackfly, and will also consume mites, scale insects, and small caterpillars. The ladybug is a very familiar beetle with various colored markings, while its larvae are initially small and spidery, growing up to 17 millimeters (mm) long. The larvae have a tapering segmented gray/black body with orange/yellow markings nettles in the garden and by leaving hollow stems and some plant debris over–winter so that they can hibernate over winter.

Hoverflies, resembling slightly darker bees or wasps, have characteristic hovering, darting flight patterns. There are over 100 species of hoverfly, whose larvae principally feed upon greenfly, one larva devouring up to 50 a day, or 1000 in its lifetime. They also eat fruit tree spider mites and small caterpillars. Adults feed on nectar and pollen, which they require for egg production. Eggs are minute (1 mm), pale yellow-white, and laid singly near greenfly colonies. Larvae are 8–17 mm long, disguised to resemble bird droppings; they are legless and have no distinct head. Therefore, they are semi–transparent with a range of colors from green, white, brown, and black. Hoverflies can be encouraged by growing attractant flowers such as the poached eggplant (Limnanthes douglasii), marigolds, or phacelia throughout the growing season.

Dragonflies are important predators of mosquitoes, both in the water, where the dragonfly naiads eat mosquito larvae, and in the air, where adult dragonflies capture and eat adult mosquitoes. Community–wide mosquito control programs that spray adult mosquitoes also kill dragonflies, thus removing an important biocontrol agent, and can actually increase mosquito populations in the long term.

Other useful garden predators include lacewings, pirate bugs, rove and ground beetles, aphid midge, centipedes, as well as larger fauna such as frogs, toads, lizards, hedgehogs, slow–worms, and birds. Cats and rat terriers kill field mice, rats, june bugs, and birds. Dogs chase away many types of pest animals. Dachshunds are bred specifically to fit inside tunnels underground to kill badgers.

Parasitoidal insects

Most insect parasitoids are wasps or flies. For example, the parasitoid Gonatocerus ashmeadi (Hymenoptera: Mymaridae) has been introduced to control the glassy-winged sharpshooter Homalodisca vitripennis (Hemipterae: Cicadellidae) in French Polynesia and has successfully controlled about 95 percent of the pest density (Hoddle et al. 2006). Parasitiods comprise a diverse range of insects that lay their eggs on or in the body of an insect host, which is then used as a food for developing larvae. Parasitic wasps take much longer than predators to consume their victims, for if the larvae were to eat too fast they would run out of food before they became adults. Such parasites are very useful in the organic garden, for they are very efficient hunters, always at work searching for pest invaders. As adults, they require high–energy fuel as they fly from place to place, and feed upon nectar, pollen and sap, therefore planting plenty of flowering plants, particularly buckwheat, umbellifers, and composites will encourage their presence.

Four of the most important groups are:

- Ichneumonid wasps: (5–10 mm) Prey mainly on caterpillars of butterflies and moths.
- Braconid wasps: Tiny wasps (up to 5 mm) attack caterpillars and a wide range of other insects including greenfly. It is a common parasite of the cabbage white caterpillar, seen as clusters of sulphur yellow cocoons bursting from collapsed caterpillar skin.
- Chalcid wasps: Among the smallest of insects (<3 mm). It parasitizes eggs/larvae of greenfly, whitefly, cabbage caterpillars, scale insects, and strawberry tortrix moth.
- Tachinid flies: Parasitize a wide range of insects including caterpillars, adult and larval beetles, true bugs, and others.

Parasitic nematodes

Nine families of nematodes (Allantone-matidae, Diplogasteridae, Heterorhabditidae, Mermithidae, Neotylenchidae, Rhabditidae, Sphaerulariidae, Steinernematidae, and Tetradonematidae) include species that attack insects and kill or sterilize them, or alter their development (UN–LN 2003). In addition to insects, nematodes can parasitize spiders, leeches, crustaceans and mollusks. An excellent example of a situation in which a nematode may replace chemicals for control of an insect is the black vine weevil, Otiorhynchus sulcatus, in cranberries. Uses of chemical insecticides on cranberry either are restricted or have not provided adequate control of black vine weevil larvae. Heterorhabditis bacteriophora NC strain was applied, and it provided more than 70 percent control soon after treatment and was still providing that same level of control a year later (Shanks 1990).

Many nematode–based products are currently available. They are formulated from various species of Steinernema and Heterorhabditis. Some of the products found in various countries are ORTHO Bio–Safe, BioVector, Sanoplant, Boden-Ntitzlinge, Helix, Otinem, Nemasys, and so forth (Smart 1995). A fairly recent development in the control of slugs is the introduction of "Nemaslug," a microscopic nematode (Phasmarhabditis hermaphrodita) that will seek out and parasitize slugs, reproducing inside them and killing them. The nematode is applied by watering onto moist soil, and gives protection for up to six weeks in optimum conditions, though is mainly effective with small and young slugs under the soil surface.

Plants to regulate insect pests

Choosing a diverse range of plants for the garden can help to regulate pests in a variety of ways, including;

- Masking the crop plants from pests, depending on the proximity of the companion or intercrop.
- Producing olfactory inhibitors, odors that confuse and deter pests.
- Acting as trap plants by providing an alluring food that entices pests away from crops.
- Serving as nursery plants, providing breeding grounds for beneficial insects.
- Providing an alternative habitat, usually in a form of a shelterbelt, hedgerow, or beetle bank, where beneficial insects can live and reproduce. Nectar–rich plants that bloom for long periods are especially good, as many beneficials are nectivorous during the adult stage, but parasitic or predatory as larvae. A good example of this is the soldier beetle, which is frequently found on flowers as an adult, but whose larvae eat aphids, caterpillars, grasshopper eggs, and other beetles.

Pathogens to be used as biopesticides

Various bacterial species are widely used in controlling the pests as well as weeds. The best-known bacterial biological control which can be introduced in order to control butterfly caterpillars is Bacillus thuringiensis, popularly called Bt. This is available in sachets of dried spores, which are mixed with water and sprayed onto vulnerable plants such as brassicas and fruit trees. After ingestion of the bacterial preparation, the endotoxin liberated and activated in the midgut will kill the caterpillars, but leave other insects unharmed. There are strains of Bt that are effective against other insect larvae. Bt. israelensis is effective against mosquito larvae and some midges.

Viruses most frequently considered for the control of insects (usually sawflies and Lepidoptera) are the occluded viruses, namely NPV, cytoplasmic polyhedrosis (CPV), granulosis (GV), and entomopox viruses (EPN). They do not infect vertebrates, non-arthropod invertebrates, microorganisms, and plants. The commercial use of virus insecticides has been limited by their high specificity and slow action.

Fungi are pathogenic agents to various organisms including the pests and the weeds. This feature is intensively used in biocontrol. The entomopathogenic fungi, like Metarhizium anisopliae, Beauveria bassiana, and so forth cause death to the host by the secretion of toxins. A biological control being developed for use in the treatment of plant disease is the fungus Trichoderma viride. This has been used against Dutch Elm disease, and to treat the spread of fungal and bacterial growth on tree wounds. It may also have potential as a means of combating silver leaf disease.

Significance of biological control

Biological control proves to be very successful economically, and even when the method has been less successful, it still produces a benefit-to-cost ratio of 11:1. The benefit-to-cost ratios for several successful biological controls have been found to range from 1:1 to 250:1. Further, net economic advantage for biological control without scouting vs. conventional insecticide control ranged from $ 7.43 to $ 0.12 per hectare in some places. It means that even if the yield produce under biological control be below that for insecticidal control by as much as 29.3 kilos per hectare, the biological control would not lose its economic advantage (CNR 2007).

Biological control agents are non-polluting and thus environmentally safe and acceptable. Usually they are species specific to targeted pest and weeds. The biological control discourages the use of environmentally and ecologically unsuitable chemicals, so it always leads to the establishment of natural balance. The problems of increased resistance in the pest will not arise, as both biological control agents and the pests are in complex race of evolutionary dynamism. Because of chemical resistance developed by the Colorado potato beetle (CPB), its control has been achieved by the use of bugs and beetles (Hein).

NEGATIVE RESULTS OF BIOLOGICAL CONTROL

Biological control tends to be naturally self–regulating, but as ecosystems are so complex, it is difficult to predict all the consequences of introducing a biological controlling agent (HP 2007). In some cases, biological pest control can have unforeseen negative results, that could outweigh all benefits. For example, when the mongoose was introduced to Hawaii in order to control the rat population, it predated on the endemic birds of Hawaii, especially their eggs, more often than it ate the rats. Similarly, the introduction of the cane toad to Australia 50 years ago to eradicate a beetle that was destroying sugar beet has been spreading as a pest throughout eastern and northern Australia at a rate of 35 km/22 mi a year. Since the cane toad is poisonous, it has few Australian predators to control its population (HP 2007).

Difficulties

Many of the most important pests are exotic, invasive species that severely impact agriculture, horticulture, forestry and urban environments. They tend to arrive without their co-evolved parasites, pathogens and predators, and by escaping from these, populations may soar. Importing the natural enemies of these pests may seem a logical move but this may have unintended consequences; regulations may be ineffective and there may be unanticipated effects on biodiversity, and the adoption of the techniques may prove challenging because of a lack of knowledge among farmers and growers.

Side-effects

Biological control can affect biodiversity through predation, parasitism, pathogenicity, competition, or other attacks on non-target species. An introduced control does not always target only the intended pest species; it can also target native species. In Hawaii during the 1940s parasitic wasps were introduced to control a lepidopteran pest and the wasps are still found there today. This may have a negative impact on the native ecosystem, however, host range and impacts need to be studied before declaring their impact on the environment.

Vertebrate animals tend to be generalist feeders, and seldom make good biological control agents; many of the classic cases of "biocontrol gone awry" involve vertebrates. For example, the cane toad (Rhinella marina) was intentionally introduced to Australia to control the greyback cane beetle (Dermolepida albohirtum), and other pests of sugar cane. 102 toads were obtained from Hawaii and bred in captivity to increase their numbers until they were released into the sugar cane fields of the tropic north in 1935. It was later discovered that the toads could not jump very high and so were unable to eat the cane beetles which stayed up on the upper stalks of the cane plants. However the toad thrived by feeding on other insects and it soon spread very rapidly; it took over native amphibian habitat and brought foreign disease to native toads and frogs, dramatically reducing their populations. Also when it is threatened or handled, the cane toad releases poison from parotoid glands on its shoulders; native Australian species such as goannas, tiger snakes, dingos and northern quolls that attempted to eat the toad were harmed or killed. However, there has been some recent evidence that native predators are adapting, both physiologically and through changing their behaviour, so in the long run, their populations may recover.

Rhinocyllus conicus, a seed-feeding weevil, was introduced to North America to control exotic musk thistle (Carduus nutans) and Canadian thistle (Cirsium arvense). However the weevil also attacks native thistles, harming such species as the endemic Platte thistle (Cirsium neomexicanum) by selecting larger plants (which reduced the gene pool), reducing seed production and ultimately threatening the species' survival.

The small Asian mongoose (Herpestus javanicus) was introduced to Hawaii in order to control the rat population. However the mongoose was diurnal, and the rats emerged at night; so it preyed on the endemic birds of Hawaii, especially their eggs, more often than it ate the rats, and now both rats and mongooses threaten the birds. This introduction was undertaken without understanding the consequences of such an action. No regulations existed at the time, and more careful evaluation should prevent such releases now.

The sturdy and prolific eastern mosquitofish (Gambusia holbrooki) is a native of the southeastern United States and was introduced around the world in the 1930s and 40s to feed on mosquito larvae and thus combat malaria. However, it has thrived at the expense of local species, causing a decline of endemic fish and frogs through competition for food resources, as well as through eating their eggs and larvae. In Australia, the mosquitofish is the subject of discussion as to how best to control it; in 1989 it was said that "biological population control is well beyond present capabilities", and this remains the position.

GROWER EDUCATION

A potential obstacle to the adoption of biological pest control measures is growers sticking to the familiar use of pesticides. However, pesticides have a variety of undesired effects, including the development of resistance among pests, and the destruction of natural enemies; these may in turn enable outbreaks of pests of other species than the ones originally targeted, and on crops at a distance from those treated with pesticides. One method of increasing grower adoption of biocontrol methods involves letting them learn by doing, for example showing them simple field experiments, enabling them to observe the live predation of pests, or demonstrations of parasitised pests. In the Philippines, early season sprays against leaf folder caterpillars were common practice, but growers were asked to follow a 'rule of thumb' of not spraying against leaf folders for the first 30 days after transplanting; participation in this resulted in a reduction of insecticide use by 1/3 and a change in grower perception of insecticide use.

11

TRAP CROPPING

Trap crops are the plant stands that are grown to attract insects or other organisms to protect target crops from pest attack. Protection may be achieved either by preventing the pests from reaching the crop or by concentrating them in certain part of the field where they can economically be destroyed.

The principle of trap cropping rests on the fact that virtually all pests show a distinct preference to certain crop stage. Manipulation of stand in time and space so that attractive host plants are offered at critical time in pests and the crop phenology leads to the concentration of the pests at the desired site, the crop.

Farmers are being motivated to utilize trap cropping because of the difficulties in cropping with the pest situations in other ways. Some times the cost of chemical pesticides and the number of treatments required is so high that more economical ways have to be developed, additionally, the pests have often evolved resistance to commonly used pesticides, which requires some alternative control strategies.

Further, motivations to use trap cropping are economical and environmental benefits are often associated with this strategy. Besides, its potential role in improving the environmental soundness, trap cropping techniques may have special preference of conventional agriculture to sustainable farming in developing countries.

Additionally, the increasing sector of organic farming also could exploit this strategy of pest control. Yet another function of trap crop is their use of attracting natural enemies of pest insects to the fields and concentrating them there to enhance naturally occurring biological control'.

The essential features of the trap cropping are that the trap crop must be attractive to the pest then the main crop, it should occupy small area as far as possible and it should be established an early or later or along with the main crop.

The important trap crops commonly used in pest management included, bhendi/okra in cotton to trap bollworms at the ratio of 1:10 and marigold at the boarder of the field. Sesamum is commonly being used as trap crop to attract Diamondback moth in both cabbage and cauliflower. Two rows of sesamum for every 25 rows of cabbage or cauliflower can be planted to trap the pest. In groundnut, castor or sunflower can be used to attract leaf eating caterpillar on the boarder of the field. In tomato marigold or cucumber is commonly used as trap crop for every 15 rows of the main crop to attract tomato fruit borer. In case of field beans, chrysanthamum acts as a trap crop against leaf minor. Marigold is a potential trap crop in potato and rice against nematodes and snails, respectively. To trap corn stalk borer in maize sorghum has been exploited as trap crop. Bihar hairy caterpillar in cowpea can be trapped by planting Gingelly.

Trap cropping has indicated a great benefit interms of economic returns on an average of 10-30 per cent increase in net profits mainly resulting from reduced insecticide use and pest attack. Trap cropping is a useful strategy in the management of several pests in various cropping systems. It offers significant economic and environmental benefits and it can successfully integrated with cultural, biological and chemical control methods.

Most gardeners have experienced the disappointment of carefully raising a vegetable crop only to have it damaged or destroyed by an invasion of pests such as slugs, aphids or other bugs. The traditional advice in such situations is to heavily spray crops with pesticides but many of us prefer to use nature's own organic controls for the food we are going to eat. Trap cropping is one of those valuable organic techniques and is regularly used by organic farmers yet few gardeners are aware of the benefits it offers.

A trap crop, also known as a sacrificial crop, is a plant that you add to your garden to attract pests away from the main crops you are growing. The reasoning is this: just as many children will choose ice-cream over a plate of vegetables, likewise most garden pests have preferences for what they like to live on. By planting rows of a trap crop (the 'ice-cream') near to your vegetables the pests will be attracted to the trap crop and will usually leave your main crops alone. You don't harvest anything from your trap crop ñ it is just there to keep the pests off the plants you want to do well.

There are several factors which need to be taken into consideration when planting a trap crop:

- Different insects prefer different trap crops: Which trap crop you choose depends on the pests you are trying to trap. If the plant isn't sufficiently attractive to the pest then it won't be any use so it is vital to pick the right ones. Often this is a matter for experimentation coupled with observation of what the pests go for in your garden.
- Layout of the trap crop: For some insects it is sufficient to plant the trap crop around the border of your growing area. Others are harder to stop and it may be necessary to interplant them to draw them off the main crops. Quantities will depend on the insect you are trying to deter but farmers usually set aside something in the region of 20% of the main crop area for the trap crop. Smaller gardens growing a variety of vegetables will often need less than this.
- Timing: Most insect invasions happen at a specific time of year. For example, I almost always get an aphid invasion in late May or early June. It is important to have the trap crop already well established by the time pests arrive.
- Beneficial Insects: Trap crops are just one part of good organic pest control and need to be balanced with adequate companion planting of flowers to attract beneficial insects such as lacewings and ladybugs which feed on the pests.

A common concern with trap crops is that they could just act as breeding grounds for the pest which will then move on to the main crops. This can happen but several factors usually prevent things getting out of hand:

- When pests increase so do their predators. The pests may be building up on the trap crop but as long as you have built in some companion plants there will usually be a hungry population of beneficial insects to start eating them. Often this controls the pests sufficiently and no further action is necessary.
- When a trap crop becomes overrun with a pest you can remove it or thin it out. The pests are then disposed of with the plant on the compost heap or somewhere further away from the garden.
- Alternatively choosing a trap crop with a long growing season or making more than one sowing of the trap crop can keep the pests at bay until the main harvest is complete.
- Nasturtiums are very attractive to aphids (blackfly, greenfly, whitefly) which will often completely cover the stems. Expect to see ants 'farming' the aphids for the honeydew they release.

- Nettles also attract aphids and because they do this early on in the season they are often followed by beneficial insects such as ladybugs.
- Chervil is said to be very attractive to slugs.
- French Marigold is variously reported to attract slugs, thrips and nematodes.
- Radish is said to attract flea beetle and root fly away from cabbages although various other brassicas can be used as trap crops too so it is best to experiment with Chinese cabbage and collards as well.

There is an excellent summary of the various trap crop combinations supported by research together with the pests they control on the German-based Pesticide Action Network website (in English).

Decoy Trap Plants for the Home Garden

While articles on how to use trap crops abound, specific trap crop info is scarce, particularly for the smaller home garden. The following list is compiled to give the home gardener ideas for using decoy plants, but is by no means complete:

Plant	Attracts
Dill	Tomato hornworms
Millet	Squash bugs
Amaranth	Cucumber beetle
Sorghum	Corn earworms
Radishes	Flea beetles, Harlequin bugs, and Cabbage maggots
Collards	Cabbage worm
Nasturtiums	Aphids
Sunflowers	Stinkbugs
Okra	Tomato aphids
Zinnias	Japanese beetles
Mustard	Harlequin bugs
Marigolds	Root nematodes
Eggplant	Colorado potato beetles

Fig 11.1

In addition to using decoy plants such as the above, other plants can be used to repel invading insects. Chives will repel aphids. Basil repels tomato hornworms. Tomatoes repel asparagus beetles. Marigolds are not only detrimental to nematodes; they repel cabbage moths, too.

Will using decoy plants completely eliminate your insect pest problem? Probably not, but if reducing the amount of pesticides you use in your garden or increasing yields without pesticides is your goal, learning how to use trap crops may bring you a little closer to your ideal garden.

BENEFITS OF TRAP CROPPING

By using trap cropping, farmers can reduce inputs (fuel, labor, time, and insecticides) resulting in increased income while protecting pollinators and other beneficial insects. As an example, one producer from St. Peters has not sprayed any insecticides to his cucurbit cash crop since 2011. He only applies a small amount of insecticide to the Blue Hubbard seedlings shortly before transplanting them to the corners of his fields. He currently sells his cucurbit produce as insecticide-free

Implementing trap cropping in gardens and small farms

With some planning, using trap crop plants can be easy and inexpensive. Blue Hubbard squash can also be used as a monitoring tool. For example, start growing Blue Hubbard squash transplants in mid- or late-April, and set one tray with Blue Hubbard squash seedlings out in the field. As soon as squash bugs (and/or cucumber beetles) are seen, then it will be time to transplant the trap crop plants to the field. The key is to transplant 2- week old Blue Hubbard seedlings (trap crop plants) to the field at the same time you sow the seeds of your cucurbit cash crop. If you grow your cash crop from transplants, then you will need to transplant the Blue Hubbard seedlings at least 2 weeks before your cash crop. The bottom line is that, to be most effective, Blue Hubbard squash plants need to be bigger than cash crop plants.

A second piece of advice is to kill the insect pests on the trap crop plants. Remember, having 1-2 squash bugs on your cash crop doesn't mean that you need to spray. Our research has demonstrated that for a small garden of 100 or so cucurbit plants, you can be successful at controlling squash bugs (and cucumber beetles) using 6-8 Blue Hubbard squash plants. The trap crop plants can be planted at the corners, at some distance (3-8 ft.) from the cash crop.

On a commercial farm, for cucurbits grown using plastic mulch and drip irrigation then we recommend you transplant 2-4 Blue Hubbard squash seedlings to both ends of each row.

Blue Hubbard squash can also be grown in large pots. Some farmers have implemented trap cropping outside high tunnels to protect cucumbers by placing potted Blue Hubbard plants treated with a systemic insecticide early in the season, with good results. If temperatures at night are expecting to drop potentially injuring the Blue Hubbard squash seedlings, then pots can be moved indoors.

Removing / killing insect pests congregating on trap crop plants

Insect pests (e.g., squash bugs) congregating on trap crop plants need to be eliminated. Otherwise, they are likely to reproduce on those plants, and then they can move to the cash crop. Not killing the squash vine borer means that the trap crop plants will succumb. Therefore, for both squash bug and squash vine borer control insecticides applied to the Blue Hubbard squash are recommended. Whether using organic or synthetic insectici des, make sure to apply them thoroughly, especially near the base of trap crop plants.

Synthetic insecticides: Foliar applications of insecticides listed in the Midwest Vegetable Production Guide for Commercial Growers for use in cucurbit crops will result in effective control of squash bugs and squash vine borer. Some producers have opted for using small amounts of systemic insecticides (examples of trade names: Admire Pro®, Alias 2F®) applied to the trap crop plants only. When applied to the roots, systemic insecticides are absorbed by plant tissue, producing plants that will be toxic to insects that feed on them. Imidacloprid-treated plants will continue to kill insect pests for about 3-4 weeks, depending on weather conditions.

Organic insecticides: Pyrethrins-containing insecticides such as AZERA® and Pyganic® represent good options to kill cucurbit pests. Because squash bugs are hard to kill, then use the high label rate if you are targeting the adult stage. If sprays are targeting the nymphs (immature stages), then a low or medium application rate will be effective. Timing of application is important particularly in the case of the squash vine borer, as again, once the caterpillar hatch from eggs and bores into the stem, no control method will be effective.

During bloom, sprays should be made early in the morning or later in the evening to reduce the effect of insecticides on honeybees or other pollinators. ALWAYS follow label instructions and safety procedures, and check to make sure the chemical you are using is registered for use in your crop. Mention of trade names or commercial products does not constitute endorsement or recommendation for use. If you are an organic certified producer make sure that the substance, including its brand name and formulation, is listed in your organic system plan, reviewed, and approved by your USDA-accredited certification agency.

EVALUATING THE SUCCESS OF TRAP CROPPING

Factors influencing the efficacy of trap cropping include (1) having Blue Hubbard squash under water / nutrient stress, (2) having cash crop plants bigger in size than the trap crop plants, and (3) not removing the pest from trap crop plants. Provide the Blue Hubbard squash with enough water and nutrients - in the same way you manage the cash crop.

One way to determine the success of trap cropping is by scouting both the trap crops and the cash crop at least once a week. For squash bugs, apply insecticide to the cash crop if cash crop plants are in the seedling stage, squash bugs are present, and some wilting is observed. At the early flowering stage, cash crop plants can tolerate an average of 1 egg mass per plant.

Operation

Recent studies on host-plant finding have shown that flying pests are far less successful if their host-plants are surrounded by any other plant, or even "decoy-plants" made of green plastic, cardboard or any other green material. The host-plant finding process occurs in three phases.

The first phase is stimulation by odours characteristic to the host-plant. This induces the insect to try to land on the plant it seeks. But insects avoid landing on brown (bare) soil. So if only the host-plant is present, the insects will quasi-systematically find it by landing on the only green thing around. This is called an "appropriate landing". When it does an "inappropriate landing", it flies off to any other nearby patch of green. It eventually leaves the area if there are too many "inappropriate" landings.

The second phase of host-plant finding is for the insect to make short flights from leaf to leaf to assess the plant's overall suitability. The number of leaf-to-leaf flights varies according to the insect species and to the host-plant stimulus received from each leaf. But the insect must accumulate sufficient stimuli from the host-plant to lay eggs; so it must make a certain number of consecutive "appropriate" landings. Hence if it makes an "inappropriate landing", the assessment of that plant is negative and the insect must start the process anew.

Thus it was shown that clover used as a ground cover had the same disruptive effect on eight pest species from four insect orders. An experiment showed that 36% of cabbage root flies laid eggs beside cabbages growing in bare soil (which resulted in no crop), compared with only 7% beside cabbages growing in clover (which allowed a good crop). Also that simple decoys made of green card disrupted appropriate landings just as well as did the live ground cover.

Alfalfa

Alfalfa is a perennial flowering plant in the pea family Fabaceae cultivated as an important forage crop in many countries around the world. It is used for grazing, hay, and silage, as well as a green manure and cover crop. The name alfalfa is used in North America. The name lucerne is the more commonly used name in the United Kingdom, South Africa, Australia, and New Zealand. The plant superficially resembles clover (a cousin in the same family), especially while young, when trifoliate leaves comprising round leaflets predominate. Later in maturity, leaflets are elongated. It has clusters of small purple flowers followed by fruits spiralled in 2 to 3 turns containing 10–20 seeds. Alfalfa is native to warmer temperate climates. It has been cultivated as livestock fodder since at least the era of the ancient Greeks and Romans. Alfalfa sprouts are a common ingredient in dishes made in South Indian cuisine.Ecology

Alfalfa is a perennial forage legume which normally lives four to eight years, but can live more than 20 years, depending on variety and climate. The plant grows to a height of up to 1 m (3.3 ft), and has a deep root system, sometimes growing to a depth of more than 15 m (49 ft) to reach groundwater. Typically the root system grows to a depth of 2–3 metres depending on subsoil constraints. This depth of root system, and perenniality of crowns that store carbohydrates as an energy reserve, make it very resilient, especially to droughts. Alfalfa is more drought-hardy than drought-tolerant and the persistence of the plant also depends on the management of the stand. It has a tetraploid genome.

Alfalfa is a small-seeded crop, and has a slowly growing seedling, but after several months of establishment, forms a tough "crown" at the top of the root system. This crown contains shoot buds that enable alfalfa to regrow many times after being grazed or harvested; however, overgrazing of the buds will reduce the new leaves on offer to the grazing animal.

This plant exhibits autotoxicity, which means it is difficult for alfalfa seed to grow in existing stands of alfalfa. Therefore, alfalfa fields are recommended to be rotated with other species (for example, corn or wheat) before reseeding.

Culture

Alfalfa is widely grown throughout the world as forage for cattle, and is most often harvested as hay, but can also be made into silage, grazed, or fed as greenchop. Alfalfa usually has the highest feeding value of all common hay crops. It is used less frequently as pasture. When grown on soils where it is well-adapted, alfalfa is often the highest-yielding forage plant, but its primary benefit is the combination of high yield per hectare and high nutritional quality. Its primary use is as feed for high-producing dairy cows, because of its high protein content and highly digestible fiber, and secondarily for beef cattle, horses, sheep, and goats. Alfalfa hay is a widely used protein and fiber source for meat rabbits. In poultry diets, dehydrated alfalfa and alfalfa leaf concentrates are used for pigmenting eggs and meat, because of their high content in carotenoids, which are efficient for colouring egg yolk and body lipids. Humans also eat alfalfa sprouts in salads and sandwiches.Dehydrated alfalfa leaf is commercially available as a dietary supplement in several forms, such as tablets, powders and tea. Fresh alfalfa can cause bloating in livestock, so care must be taken with livestock grazing on alfalfa because of this hazard.

Like other legumes, its root nodules contain bacteria, Sinorhizobium meliloti, with the ability to fix nitrogen, producing a high-protein feed regardless of available nitrogen in the soil. Its nitrogen-fixing ability (which increases soil nitrogen) and its use as an animal feed greatly improve agricultural efficiency.

Alfalfa can be sown in spring or fall, and does best on well-drained soils with a neutral pH of 6.8 – 7.5. Alfalfa requires sustained levels of potassium and phosphorus to grow well. It is moderately sensitive to salt levels in both the soil and irrigation water, although it continues to be grown in the arid southwestern United States, where salinity is an emerging issue. Soils low in fertility should be fertilized with manure or a chemical fertilizer, but correction of pH is particularly important. Usually a seeding rate of 13 – 20 kg/hectare (12 – 25 lb/acre) is recommended, with differences based upon region, soil type, and seeding method. A nurse crop is sometimes used, particularly for spring plantings, to reduce weed problems and soil erosion, but can lead to competition for light, water, and nutrients.

In most climates, alfalfa is cut three to four times a year, but it can be harvested up to 12 times per year in Arizona and southern California. Total yields are typically around eight tonnes per hectare (four short tons per acre) in temperate environments, but yields have been recorded up to 20 t/ha (16 ts per acre). Yields vary with region, weather, and the crop's stage of maturity when cut. Later cuttings improve yield, but with reduced nutritional content.

Beneficial insects

Alfalfa is considered an insectary, a place where insects are reared, and has been proposed as helpful to other crops, such as cotton, if the two are interplanted, because the alfalfa harbours predatory and parasitic insects that would protect the other crop. Harvesting the alfalfa by mowing the entire crop area destroys the insect population, but this can be avoided by mowing in strips so that part of the growth remains.

Pests and diseases

Like most plants, alfalfa can be attacked by various pests and pathogens. Diseases often have subtle symptoms which are easily misdiagnosed and can affect leaves, roots, and stems.

Some pests, such as the alfalfa weevil, aphids, armyworms, and the potato leafhopper, can reduce alfalfa yields dramatically, particularly with the second cutting when weather is warmest. Registered insecticides or chemical controls are sometimes used to prevent this and labels will specify the withholding period before the forage crop can be grazed or cut for hay or silage. Alfalfa is also susceptible to root rots, including Phytophthora, Rhizoctonia, and Texas root rot.

Harvesting

When alfalfa is to be used as hay, it is usually cut and baled. Loose haystacks are still used in some areas, but bales are easier for use in transportation, storage, and feed. Ideally, the first cutting should be taken at the bud stage, and the subsequent cuttings just as the field is beginning to flower, or one-tenth bloom because carbohydrates are at their highest. When using farm equipment rather than hand-harvesting, a swather cuts the alfalfa and arranges it in windrows. In areas where the alfalfa does not immediately dry out on its own, a machine known as a mower-conditioner is used to cut the hay. The mower-conditioner has a set of rollers or flails that crimp and break the stems as they pass through the mower, making the alfalfa dry faster. After the alfalfa has dried, a tractor pulling a baler collects the hay into bales.

Several types of bales are commonly used for alfalfa. For small animals and individual horses, the alfalfa is baled into small, two-string bales, commonly named by the strands of string used to wrap it. Other bale sizes are three-string, and so on up to half-ton (six-string) "square" bales – actually rectangular, and typically about 40 x 45 x 100 cm (14 x 18 x 38 in). Small square bales weigh from 25 to 30 kg (55 to 66 lb) depending on moisture, and can be easily hand separated into "flakes". Cattle ranches use large round bales, typically 1.4 to 1.8 m (4.6 to 5.9 ft) in diameter and weighing from 500 to 1,000 kg (1,100 to 2,200 lb). These bales can be placed in stable stacks or in large feeders for herds of horses or unrolled on the ground for large herds of cattle. The bales can be loaded and stacked with a tractor using a spike, known as a bale spear, that pierces the center of the bale, or they can be handled with a grapple (claw) on the tractor's front-end loader. A more recent innovation is large "square" bales, roughly the same proportions as the small squares, but much larger. The bale size was set so stacks would fit perfectly on a large flatbed truck. These are more common in the western United States.

When used as feed for dairy cattle, alfalfa is often made into haylage by a process known as ensiling. Rather than being dried to make dry hay, the alfalfa is chopped finely and fermented in silos, trenches, or bags, where the oxygen supply can be limited to promote fermentation. The anaerobic fermentation of alfalfa allows it to retain high nutrient levels similar to those of fresh forage, and is also more palatable to dairy cattle than dry hay. In many cases, alfalfa silage is inoculated with different strains of microorganisms to improve the fermentation quality and aerobic stability of the silage.

WORLDWIDE PRODUCTION

During the early 2000s, alfalfa was the most cultivated forage legume in the world. Worldwide production was around 436 million tons in 2006. In 2009, alfalfa was grown on approximately 30 million hectares (74,000,000 acres) worldwide; of this North America produced 41% (11.9 million hectares; 29,000,000 acres), Europe produced 25% (7.12 million hectares; 17,600,000 acres), South America produced 23% (7 million hectares; 17,000,000 acres), Asia produced 8% (2.23 million hectares; 5,500,000 acres), and Africa and Oceania produced the remainder. The US was the largest alfalfa producer in the world by area in 2009, with 9 million hectares (22,000,000 acres), but considerable production area is found in Argentina (6.9 million hectares; 17,000,000 acres), Canada (2 million hectares; 4,900,000 acres), Russia (1.8 million hectares; 4,400,000 acres), Italy (1.3 million hectares; 3,200,000 acres), and China (1.3 million hectares; 3,200,000 acres).

United States

In the United States in 2014, the leading alfalfa-growing states were California, Idaho, and Montana. Alfalfa is predominantly grown in the northern and western United States; it can be grown in the southeastern United States, but leaf and root diseases, poor soils, and a lack of well-adapted varieties are often limitations.

Alfalfa and bees

Alfalfa seed production requires the presence of pollinators when the fields of alfalfa are in bloom. Alfalfa pollination is somewhat problematic, however, because western honey bees, the most commonly used pollinator, are less than ideal for this purpose; the pollen-carrying keel of the alfalfa flower trips and strikes pollinating bees on the head, which helps transfer the pollen to the foraging bee. Western honey bees, however, do not like being struck in the head repeatedly and learn to defeat this action by drawing nectar from the side of the flower. The bees thus collect the nectar, but carry no pollen, so do not pollinate the next flower they visit. Because older, experienced bees do not pollinate alfalfa well, most pollination is accomplished by young bees that have not yet learned the trick of robbing the flower without tripping the head-knocking keel. When western honey bees are used to pollinate alfalfa, the beekeeper stocks the field at a very high rate to maximize the number of young bees. Western honey bee colonies may suffer protein stress when working alfalfa only, because of shortage of one of the amino acids comprising the pollen protein, isoleucine. Today, the alfalfa leafcutter bee is increasingly used to circumvent these problems. As a solitary but gregarious bee species, it does not build colonies or store honey, but is a very efficient pollinator of alfalfa flowers. Nesting is in individual tunnels in wooden or plastic material, supplied by the alfalfa seed growers. The leafcutter bees are used in the Pacific Northwest, while western honeybees dominate in California alfalfa seed production.

A smaller amount of alfalfa produced for seed is pollinated by the alkali bee, mostly in the northwestern United States. It is cultured in special beds near the fields. These bees also have their own problems. They are not portable like honey bees, and when fields are planted in new areas, the bees take several seasons to build up. Honey bees are still trucked to many of the fields at bloom time.

B. affinis is important to the agricultural industry, as well as for the pollination of alfalfa. It is known that members of this species pollinate up to 65 different species of plants, and it is the primary pollinator of key dietary crops, such as cranberries, plums, apples, onions, and alfalfa.

Megachile rotundata is another species that pollinates the alfalfa plant. M. rotundata was unintentionally introduced into the United States during the 1940s, and its management as a pollinator of alfalfa has led to a three-fold increase in seed production in the U.S. The synchronous emergence of the adult bees of this species during alfalfa blooming period in combination with such behaviors as gregarious nesting, and utilization of leaves and nesting materials that have been mass-produced by humans provide positive benefits for the use of these bees in pollinating alfalfa.

PESTS AND DISEASES OF ROSES

Roses (Rosa sp.) are susceptible to a number of pests, diseases and disorders. A large number of the problems affecting roses are seasonal and climatic. Certain varieties of roses are naturally more resistant or immune than others to certain pests and diseases. Cultivation requirements of individual rose species and cultivars, when observed, often assist in the prevention of certain pests, diseases and disorders.Pests

Insects that affect roses are often considered pests.

- Aphids: (Greenfly) (Order Hemiptera Family Aphididae) Macrosiphum rosae – Likely to be found on new shoots and buds, aphids are soft bodied insects 1-2mm long. Often green but occasionally light-brown, and sometimes with wings, they may cover (in a colony) the complete growing tip of the plant. Aphids are most active in spring and summer and multiply at a prodigious rate feeding on the sap of the plant by piercing the plant cells via a proboscis. In large quantities they may seriously retard the growth of the plant and ruin buds. They are particularly damaging to the new shoots with subsequent damage to the emerging leaves which become malformed with much the same appearance as leaf-curl in peaches.
- Two-spotted mite (Spider-mites or Red spider mite) (Order Acari: Family Tetranychidae) Tetranychus urticae – Previously known as red-spider mite these arachnids prefer the underside of leaves and are difficult to see with an unaided eye. Evidence of their presence is silvering of leaves where the mites have destroyed individual leaf cells. Fine webbing and eggs on the undersides of leaves is further evidence of the presence of Tetranychus urticae.
- Thrips (Order Thysanoptera) – Thrips are slim-winged insects 1mm in length, resembling fine slivers of wood or rice. Preferring light-coloured blooms and often appearing in plague numbers, flowers are often left looking scarred, warped, and lustreless.
- Rose Slugs (Rose Sawflies)- Sawflies are non-stinging wasps (Hymenoptera) in the suborder Symphyta, not flies (Diptera). They lay eggs in plant leaves or stems with a saw-like ovipositor. There are three species that commonly cause damage to wild or cultivated roses: The Bristly Roseslug (Cladius difformis) is found in Europe, Siberia, and many areas of North America. The larvae is pale green, up to 16 millimetres (0.6 in) long, and covered with hairlike bristles all over its body. It looks like a caterpillar but that term strictly speaking only applies to the larvae of moths and butterflies. It skeletonizes the underside of leaves, with several generations per year. The European Roseslug (Endelomyia aethiops) is found in North America as well as Europe. The larvae is more slug-like (but not slimy), up to 13 millimetres (0.5 in), and skeletonizes the upper surface of leaves with only one generation per year. The Curled Roseslug (Allantus cinctus) larvae is pastel green on the back, marked on the thorax and abdomen with white dots, and up to 19 millimetres (0.7 in) long. It frequently coils up like a snake. After skeletonizing entire leaves except the main veins, it pupates in the pith of canes, with up to two generations per year in North America.

- Caterpillars (Order Lepidoptera) See also List of Lepidoptera that feed on roses – The tortryx (tortrix) moth Lozotaenia forsterana is a prominent pest of roses, although not the sole pest. The caterpillars are green, up to 15mm long, and can be found boring into buds or within curled leaves. When disturbed the caterpillars move swiftly, dropping to the ground on a fine thread. Damage is chewn leaves and flowers and buds with "shot holes".
- Curculio beetles (Family Curculionidae)
- Japanese beetles (Popillia japonica) - This species, introduced to North America in 1912, is now an endemic pest in the eastern United States. Adult beetles emerge from the ground in early summer and join into swarms for four to six weeks, devouring blooms and skeletonizing foliage on roses and many other garden plants. Japanese beetles can be partially controlled, albeit slowly, by spreading milky spore bacillus on the lawn areas where the larvae live. The popular pheromone-baited traps may do more harm than good by attracting beetles from a wide area.
- Scale insects (Order Hemiptera)

Cottony cushion scale (Order Hemiptera : Family Coccoidea) Icerya purchasi – This scale infests twigs and branches. The mature female is oval in shape, reddish-brown with black hairs, 5 mm long. When mature the insect remains stationary and produces an egg sac in grooves, by extrusion, in the body which encases hundreds of red eggs. The insect causes little damage but produces copious honeydew (frass) that can cause damaging sooty mould.

California red scale (Order Hemiptera : Family Coccoidea) Aonidiella aurantii – A hard scale, orange to orange-pink, the female covering being less than 1.5mm across. Often in plague numbers this scale infests upper surfaces of foliage causing yellowing, leaf fall, and twig and branch dieback. Serious infestations can cause plant death.

Rose scale (Order Hemiptera : Family Coccoidea) Aulacaspis rosae – Mainly found on the stems and branches of the plant, lack of control will allow the pest to spread to flower stalks and petioles. At this point the plant would be stunted, spindly and with a white, flaky crust of scales on the bark. Female Aulacaspis rosae may live for 1 year and may lay 80 eggs each with several overlapping generations living within milliimetres of the original parent.

- Leaf Cutting Bee (Order Hymenoptera : Family Megachilidae) Megachile spp. – Leafcutter bees are 6-16mm long and mostly black with bands of light-coloured hair. They chew pieces from the edges of leaves. The pieces are regular in shape, circular or oval. Damage is not often significant.

- Nematodes (Eelworms)(Order Tylenchida: Family Heteroderidae)

Root-knot nematode Meloidogyne spp.

See - Root-knot nematode - symptoms of Meloidogyne infestation in roses is stunting, slow-growth, pale green leaves and wilting in mild weather.

- Rose chafer (Order Coleoptera : Family Scarabaeidae) Macrodactylus subspinosus – The rose chafer is common to North America, and emerges in late May to mid June. These beetles form aggregations and feed upon foliage, buds, flowers, and fruit of roses and other ornamentals.
- Metallic flea-beetles (Order Coleoptera: Family Chrysomelidae) Altica spp. – The small, shiny and metallic Altica beetles have thickened hindlegs adapted to jumping, similar to fleas. The insects are 3mm long and chew holes of irregular shapes in young leaves and buds. As the leaves enlarge so do the holes.

DISEASES

Fungal, bacterial and viral diseases that affect roses:

Bacterial diseases

- Crown gall rot (Class Alpha Proteobacteria: Family Rhizobiaceae) Agrobacterium rhizogenes – This disease is characterized by large lumps at the base of the plant stem or on roots. Galls may appear higher on stems as the disease progresses. Galls are soft compared to surrounding plant tissues. The pathogenic bacteria enter the plant via a wound. If the disease affects the plant whilst it is young the plant may be affected to the degree where it will not produce blooms. All affected plants wilt readily and grow poorly.

Fungal diseases

- Black Spot (Class Leotiomycetes: Family Helotiales) Diplocarpon rosae syn. Marssonina rosae – Marssonina rosae causes black spotson leaves. The spots, which may be as much as 12mm across, are generally circular and have an irregular edge often with a yellow halo. Leaves frequently turn yellow and fall early. Sometimes new leaves are produced, and these may also become affected. Continual defoliation will cause weakness, dieback or death of the plant. Some very susceptible species may have stems affected with a considerable reduction in plant vigour.

- Powdery Mildew Podosphaera pannosa – P. pannosa produces a very fine, powdery coating on the surface of buds and leaves. Significant cases have stems and particularly thorns, infected. Attacks on young leaves and buds will cause deformity with retardation of growth. Infected buds will fail to open. The disease is likely in hot, humid weather, with fungal spores overwintering on the stems and fallen leaves.
- Downy mildew (Class Oomycetes : Family Peronosporaceae) Peronospora sparsa – Peronospora causes purple-red to dark-brown spots on the leaves with irregular margins, however, often angular. Stems, petioles and flower stalks can split and spotted with purple marks. Buds, sepals, petals and calyces can be affected and will present purple spots. New growth affected will be deformed. The disease is spread by wind.
- Rust Phragmidium mucronatum – Rose rust appears as yellow patches on the surface of leaves, with orange pustules of sporesunderneath the leaf. The fungus is spread by wind. Affected leaves fall prior to healthy ones and plants may be defoliated in serious infections.
- Anthracnose Sphaceloma rosarum – Spots caused by this fungus originate from a point where leaves are water soaked, usually unnoticeable at first, until they turn black with a very distinct defined edge. As the spots enlarge the centre becomes gray and may fall out resulting in a shot-hole appearance. Defoliation may occur but is often not serious.
- Grey mould (Class Leotiomycetes: Family Sclerotiniceae) Botrytis cinerea – On roses grey mould is primarily a disease of the flowers and buds, leaves are infrequently attacked. Infected buds rot on the stem and infection may progress down the stem. On petals botrytis cinerea produces pink rings.
- Verticillium wilt (Class Incertae sedis: Family Verticillium) Verticillium dahliae
- Sooty moulds Alternaria spp. – Sooty mould appears as black, dry powder on leaves similar to chimney soot. Many sooty moulds grow on the honeydew (frass) produced by sap-sucking insect such as aphids and soft scales. Alternaria does no direct damage to plants but surface cover of leaves will reduce the plants capacity to photosynthesise and may create an unsatisfactory plant appearance.

- Canker Leptosphaeria coniothyrium and Cryptosporella umbrina – Cankers present as small yellowish or reddish spots on bark slowly increasing in size. Leptosphaeria coniothyrium turns brown, increases in size, and may eventually girdle the stem. The tissue within the infection begins to dry out and shrink, presenting a shriveled appearance. If the disease infects only part of the stem, growth above the canker will continue. If it girdles the stem, however, growth will cease and the stem will die.

Nematodes, parasitic

Plant-parasitic nematodes include several genera and can be detrimental to plants in different ways - from attacking plants, acting as vectors spreading plant viruses to endoparasites.

Phytoplasma, Virus and viruslike diseases

(And graft-transmissible pathogens [GTP])

- Rose mosaic – This disease is caused by a complex or viruses and is characterized by yellow patterns on the leaves. The patterns vary considerably, ranging between all-over fine blotches to patterns of lines in waves. The patterns may appear on a few or many leaves. Plants are infected by this virus at propagation using infected plant material.
- Rose wilt – Rose wilt is a complex of viruses and is referred to as dieback in some areas. The disease can be spread by vectors such as aphids. Symptoms are variable and range from stunted growth to curled young leaves. The soft tissue symptoms are more evident in spring and new leaves will reflex towards their own petioles. The affected leaves are brittle and easily fall from the plant. Fully formed leaves will 'wilt' as if the plant were water stressed.
- Rose rosette disease - This disease is caused by a new virus, rose rosette virus, that is transmitted by an eriophyid, rose leaf curl mite (Phyllocoptes fructiplilus), which inhabits the shoot tips and leaf petal bases of roses, as well as by grafting but not by seed or many other common vectors. Rose rosette was initially mistaken for a phytoplasma disease; however, heat and tetracycline treatments did not cure the disease showing that a phytoplasma is not the causal agent. Also called witches' broom of roses, it is fatal (average lifespan after infection 22 months) in the shrub Rosa multiflora, commonly found wild or as hedges (and considered a noxious weed in some places). It can also infect other rose species, such as garden rose climbers, miniatures, hybrid

teas, floribundas, and antique varieties and is capable of killing these as well. Roses are the only plants known to be susceptible. Symptoms include mosaic pattern on the leaves, malformed leaves and flowers, elongated shoots that are often red, and sometimes thorn proliferation. The distorted growth may be mistaken for herbicide damage. There is no treatment for the disease, and control is limited to controlling the vector and destroying infected plants. It is reported that the causal agent does not survive in the soil, but can survive in root fragments.

Environmental disorders

- Frost will destroy fresh growth causing stems and leaves to wilt, turn black and fall away from the plant. Timing pruning to promote growth after the threat of frost is a means to avoid frost damage.
- Salinity will present in roses as limp and light brown leaves with dry leaf margins. Soil may require testing to determine salinity levels. Symptoms will present if salinity is greater than 1200 parts per million.
- Herbicide Damage – Overspray or soil leaching of herbicidal sprays can present with several symptoms: :Prolonged exposure to overspray of glyphosate will cause yellow leaves and new leaves will be small and elongated. Hormone weed sprays (e.g. 24-D & 245-T) may cause grotesque new growth with thin twisted leaves and distorted buds. Plants may die in severe cases. Pre-emergent herbicidescontacting the plants' root system via the soil will cause yellowing foliage. Effects of soil borne herbicide may take several years to clear.

Bibliography

- Amy P. Cotton, '123 Years At The Negotiating Table And Still No Dessert? The Case In Support Of Trips Geographical Indication Protections', Chicago-Kent Law Review 2007 [Vol 82:3] 1295-1316.
- Briggs, J. & Sharp, J., 'Indigenous Knowledges and Development: A Postcolonial Caution', 25(4) Third World Quarterly, 2004, pp. 661 664. X
- Bruce A. Babcock and Roxanne Clemens, 'Geographical Indications and Property Rights: Protecting Value-Added Agricultural Products MATRIC Briefing Paper 04- MBP7, May 2004 at http://www.card.iastate.edu/publications/dbs/pdffiles/04mbp7.pdf
- Budi Suratno, 'Protection of Geographical Indications', [2004] IP Management Review, Vol. 2, pp. 87-93
- 'Protecting Geographical Indications as Collective Trademarks: The Prospects and Pitfalls' at http://www.iip.or.jp/e/e_summary/pdf/detail2005/e17_14.pdf
- AJ Soloman Raju & K Venkata Ramana, 'Traditional Preparation of a health drink Nannari Sharbat from the root extract of Decalepis hamiltonii Wight. and Arn.' Indian Journal of Natural Products and Resources, Vol 2(1), March 2011, pp. 121-124
- Alberto Almeida, 'Key Differences between Trademarks and Geographical Indications' available at http://citizenseminars. blogactiv.eu/files/2009/03/trade-marksin-pdf.pdf
- Amélie Lecoent, Emilie Vandecandelaere and Jean-Joseph Cadilhon, 'Quality linked to geographical origin and geographical indications: lessons learned from six case studies in Asia' RAP Publication 2010/04, available at http://www.fao.org/docrep/013/i1592e/i1592e00.pdf
- Amir Zahir, 'North & South – the WTO TRIPS Agreement: Intellectual Property Disputes and Conflict of Visio

- C. Sudhakar Reddy, K. N. Reddy, E. N. Murthy and V. S. Raju, 'Traditional medicinal plants in Seshachalam hills, Andhra Pradesh, India' Journal of Medicinal Plants Research Vol. 3(5), pp. 408-412, May, 2009 available at http://www.academicjournals.org/JMPR
- Daniel R. Bereskin, 'Legal Protection of Geographical Indications in Canada' at http://islandtastesensations.com/attachments/File/GI/TM-GeographicBereskin%5B1%5D.pdf
- Daphne Zografos, 'Geographical Indications & Socio-Economic Development' at http://www.iqsensato.org/wp-content/uploads/2009/02/iqsensato-wp-3-zografos-dec- 2008.pdf
- Dr. Bhamy V. Shenoy, 'Lessons Learned from Attempts to Reform India's Kerosene Subsidy' March 2010, available at http://www.iisd.org/pdf/2010/lessons_india_kerosene_subsidy.pdf
- Dr. Dwijen Rangnekar – 'Geographical Indications', A Review of Proposals at the TRIPS Council: Extending Article 23 to Products other than Wines and Spirits, June 2003 at http://www.iprsonline.org/resources/docs/Rangnekar%20-%20GI%20- %20Blue%204.pdf
- Dr. Sudhir Ravindran & Ms. Arya Mathew, 'The Protection of Geographical Indication in India – Case Study on 'Darjeeling Tea'. XI
- Felix Addor & Alexandra Grazioli, 'Geographical Indications beyond wines and spirits—A Roadmap for Better Protection for GIs', Journal of World Intellectual Property, vol. 5, No. 6, November 2002, pp. 865-897.
- Frank Thiedig and Bertil Sylvander, 'Welcome to the Club? - An Economical Approach to Geographical Indications in the European Union' at http://www.originfood.org/pdf/partners/bs7nov00.pdf
- Giovannucci, Daniele; Josling, Timothy; Kerr, William; O'Connor, Bernard and Yeung, May T.- Guide to Geographical Indications: Linking Products and Their Origins; Online at http://mpra.ub.uni-muenchen.de/27955/

INDEX

9 789392 969263

Printed at Repro India Ltd.